WILD TIDES

WILD TIDES — MEDIA INFRASTRUCTURE AND FINANCIAL CRISIS IN IRELAND

Patrick Brodie

DUKE UNIVERSITY PRESS
Durham and London

2 0 2 6

© 2026 DUKE UNIVERSITY PRESS
All rights reserved
Designed by A. Mattson Gallagher
Typeset in Arno Pro and Cronos Pro
by Westchester Publishing Services

Library of Congress Cataloging-in-Publication Data
Names: Brodie, Patrick, [date] author
Title: Wild tides : media infrastructure and financial crisis in
Ireland / Patrick Brodie.
Other titles: Media infrastructure and financial crisis in Ireland
Description: Durham : Duke University Press, 2026. | Includes
bibliographical references and index.
Identifiers: LCCN 2025031022 (print)
LCCN 2025031023 (ebook)
ISBN 9781478038542 paperback
ISBN 9781478033653 hardcover
ISBN 9781478062134 ebook
Subjects: LCSH: Mass media—Economic aspects—Ireland |
Infrastructure (Economics)—Social aspects—Ireland |
Globalization—Economic aspects—Ireland | Economic
development—Social aspects—Ireland | Ireland—Economic
conditions—21st century
Classification: LCC P96.E252 I73 2026 (print) | LCC P96.E252
(ebook) | DDC 302.2309415—dc23/eng/20251118
LC record available at https://lccn.loc.gov/2025031022
LC ebook record available at https://lccn.loc.gov/2025031023

Cover art: Valérie Anex, *The Waterways, Keshcarrigan*, 2011.
From the series *Ghost Estates*.

CONTENTS

Introduction

A Rising Tide Lifts All Boats

The combination of an increased reliance on short-term foreign liquidity sources and an increased reliance on property as an outlet for lending, in the short term, inflated asset bubbles further and, in the medium term, created the perfect storm for one of the most spectacular property and financial crashes in the history of capitalism.

Sinéad Kelly, "Light Touch Regulation: The Rise and Fall of the Irish Banking Sector"

A rising tide lifts all boats.

Seán Lemass, common usage

The economic fortunes of contemporary Ireland are intricately tethered to the fluctuations of global capital.[1] Since the late 1950s, and then especially since the early 1990s, Ireland's increasingly active participation in the world economic system has facilitated, on the surface, an emergent and remarkably prosperous capitalist society characterized by a global consumer culture, significant social progress, and the swelling presence of multinational business. The 2007–8 financial crisis left Ireland in a state of disrepair after tearing across the markets of much of the world, revealing the uneven but interdependent distribution of risk and prosperity after financial globalization. The high-water marks of finance capital could be seen dotted across all corners of the country, with unfinished housing projects, suburban developments, and other abandoned built environments materializing the generalized feelings of shock and hopelessness cutting across the population. Prosperity at the apparently invisible hands of global finance capital and its productive forces, as it turned out, was part of a materially interconnected, spatially expansive, and deeply extractive system of erosion, through which the world's largest owners of capital and a select few native offspring generated enormous floods of spectacular wealth, only to withdraw it, revealing the empty veneers of growth and success that mask its corrosive liquidity.

Wild Tides performs a study of some of these hidden mechanisms of value creation and extraction that undergird the social and economic development logics of the Irish state during and after this radically altering financial crisis, and especially its cultural and spatial dimensions through media infrastructure.[2] In the aftermath of the crisis, capital reanimated space in Ireland, the creeping fingers of capital's returning tide appearing to nourish its shores once again. But this "Celtic Phoenix" rising from the ashes, as some referred to it—or to carry the oceanic metaphor, a resurgent Celtic swell[3]—was chimeric for most people. Many had already figured out new ways to make a living, fled for more plentiful places, or simply never experienced the first "rising tide" everyone was talking about in the 1990s and 2000s. It seemed, for many, that this tide had only lifted *some* boats that remained afloat after the recession, while others had either sunk or run aground—or never been seaworthy to begin with—amid the turbulent and unpredictable tides of global capital.

Wild Tides looks at wreckage and rebuilding in the transformed cultural and political landscape of post–financial crisis Ireland until 2020. What systemic factors had differentially kept individuals, communities, and businesses afloat, sunk, or run aground? What processes and remnants could be attributed to the relatively novel Celtic Tiger, and which were continuous developments of Irish capitalism? And how did the state, communities, and individuals reconstruct lives, systems, and institutions in this environment? Central to this book's argument is that the Celtic Tiger and the subsequent financial crash were *spatial* phenomena, driven by irresponsible state intervention and (lack of) financial regulation in property and development, and that these strategies were reoriented toward private (media) infrastructure during the crisis and its aftermath. The question asked by so many is: How and why did the state double down on this strategy? It was not due only to external pressures, as this book will detail. Amid the turmoil of the crash, the state navigated the cultural and emotional tensions of a society still adjusting to its advanced position in the global economy. By using these cultural dynamics as a channel for multinational investment, however much at the behest of supranational debtors and enormously powerful corporations, the state and its partners in industry facilitated a remarkable reintroduction of financial and logistical flows through Irish territory in the post–financial crisis era—an era that we may now retroactively refer to as "the recovery."

But again, we must ask: By and through what mechanisms? This book analyzes these dynamics by focusing specifically on media and technology industries and their infrastructures of production and circulation in post–financial crisis Ireland. To do so, it looks closely at the public imaginations and popular discourses around media and culture coursing through this environment. *Wild Tides* thus locates the place of media within the spatial and circulatory logics of capitalism as its material operations cut across the landscape of contemporary Ireland. In this milieu, the state continued to gamble the country's, and by extension its media and creative industries', futures on the foreign direct investment (FDI)–driven dreams of finance, global media, and big tech,[4] through the imbrication of business policy, planning, and spatial development. At the time of the crisis, media industries in Ireland were transitioning to a model of financial and logistical facilitation for foreign corporate and transnational productions and, at the same time, the tech industry was increasingly setting its sights on the state as the site for low tax headquarters in Europe. The coalescing industries of media and tech, as they "converged" (Jenkins 2006) in the early 2010s with the rapid emergence of smart technologies, streaming, digital platforms, and data-driven marketing, were targeted by the state as a method to draw capital investment back into the country's ailing business environment, reproducing FDI-driven strategies that characterized the Celtic Tiger and its precipitous crash.

Wild Tides confronts these growth-driven logics, arguments, and strategies through media infrastructures, which at the large scale serve to materialize and naturalize turbulent but methodical extraction of value from space and culture by global capital. By focusing on the effects of media infrastructural development and experience among workers, communities, and the natural environment, I uncover both the false promises and potentially alternative futures offered by more textured engagements with where capital, in the words of political economists Sandro Mezzadra and Brett Neilson, "hits the ground" through media infrastructure. In doing so, we can see how media capital "shapes conditions of everyday life, always working in consonance or conflict with the active role of space and multifarious resistances in guiding and molding capital's operations" (Mezzadra and Neilson 2019, 22), whether at sites of media infrastructural development or struggles against it.

In media studies, through which this book attunes its analytical compass, attention has shifted over the past two decades to the material infrastructures of supposedly immaterial media of distanced communication and, more recently, on the digital platforms, data, and streaming services through which contemporary media economies function (Larkin 2008; Parks 2005, 2015a; Parks and Starosielski 2015; Plantin and Punathambekar 2019; Starosielski 2015). These studies have focused primarily on the distributive arrangements of media as their industries and economies are organized across space *via* infrastructures. But media infrastructures are also productive, not only distributive. As the sites within which powers—economic, political, social, cultural—become encoded, and imaginaries of certain kinds of existence are embedded, global media infrastructures, whether film studios, internet data centers, or the networks and supply chains that connect sites to one another, are increasingly important not only in our everyday lives and consumption of media and participation in culture. But they are also central in how states govern, collect data, and manage change and economic turbulence. The subfield of media infrastructure studies has focused on these entanglements in order to complicate fixed understandings of how media circulates through given environments, emphasizing the specificity of formal and informal infrastructures as they are experienced across diverse spaces on the ground—from surfacing the affective texture of everyday media in urban spaces, to understanding how people work in and around media industries in rural places, to answering why media infrastructures are developed where they are. *Wild Tides* builds on these understandings of infrastructure's multifaceted, sedimented, and entangled spatial existence to articulate the ways in which finance and logistics play a role in their construction of everyday life, focusing on the nexus of financialization, media, and technological infrastructure. Stubbornly emplacing these processes within their built environments of production *and* circulation, and the state and corporate finance, policy, and planning that conjure them, *Wild Tides* materializes and politicizes understandings of media's infrastructures and supply chains through their financial and logistical operations.[5]

"A Rising Tide Lifts All Boats"

Ireland, as a formerly "developing" postcolonial nation-state existing within the political boundaries of Europe, has a peculiar relationship to global economic history and modernity, especially in terms of its perceived place within the capitalist world system (Deckard 2016; O'Hearn 2001). As England's first colony and the laboratory in which many of its most violent and innovative mechanisms of territorial control were tested—from resource mapping to counterinsurgency[6]—the country has a fraught and textured history within discourses of European modernity. After protracted and complex revolutionary struggles for much of its modern history, the southern "Free State" seized independence in 1921. In the torrid aftermath, competing factions of pro- and anti-Treaty Republican forces fought a bloody and bitter civil war from 1921 to 1923. After a traumatic victory for the pro-Treaty forces, which accepted the partition between the twenty-six-county southern Free State and the six counties of "Northern Ireland," the south operated in the following years as a "dominion" of Great Britain, and finally moved to form a profoundly compromised Republic in 1937 (see McVeigh and Rolston 2021). The scars and violence of colonization were in large part inherited by the postcolonial southern state, which built a partitioned nation governed around a Catholic theocratic, socially conservative cultural nationalism, one that downplayed the "colonial question" in part due to the enduring and inconvenient reality of the occupied six counties in the north. The early days of the postcolonial state were thus characterized by strategic nation-building predicated on moving on from the colonial past and neglecting social division within and across its borders.[7]

But the transition away from colonial rule is always also an infrastructural one, and the fledgling Irish Free State was tasked with modernizing its industry and infrastructure in the absence of overtly extractive land management and paternalistic policies from Britain. While the result of a long and contested nation-building process, the mainstream history follows something like this: Under the rule of former revolutionary Eamon de Valera's Fianna Fáil party, the newly established Republic of Ireland remained neutral during World War II, focusing instead on industrializing while adapting to internal and external changes in the global economy. As the story goes, de Valera's Ireland was deeply protectionist, in part a result of the culturally Catholic insularity that constituted the "imagined" Ireland he was trying to build.[8] However, one of the primary figures in his governments through the 1950s was fellow former revolutionary Seán Lemass. Serving as the minister for industry and commerce for several tenures, most influentially prior to his election as Taoiseach from 1959 to 1966, Lemass's economic philosophy, alongside that of civil servant, economist, and governor of the Central Bank of Ireland T. K. Whitaker, espoused a cautious and experimental economic liberalism, predicated on mostly US investment. Lemass supported the establishment of the "world's first free trade zone" in Shannon, a small town in the west of Ireland, which capitalized on an existing transatlantic airport experiment and duty free zone pioneered by local entrepreneur Brendan O'Regan (see O'Connell and O'Carroll 2018). He also, significantly for the arguments of this book, presided over the establishment of Ardmore Studios in County Wicklow in 1958, a film studio partially funded by US interests to facilitate "Hollywood-style" production in the Dublin region. As Taoiseach, Lemass advocated strongly for Ireland's admission to the free market European Economic Community (EEC) (not achieved until he left office in 1971).[9]

These strategies formed the template for visions of how Ireland would be opened up to the world for business over the next decades. They were also profoundly influential for liberalization measures as implemented by postcolonial and developmental nation-states worldwide, including for much more spectacular experiments with export-processing zones implemented in China and elsewhere in the Global South (see Easterling 2014). This early liberalization period, from Lemass and Whitaker to the Celtic Tiger "boom" in the 1990s, would firmly tether Ireland's fortunes to the rolling waves of the global economy. These ties would only tighten as the tides rocked and the state solidified the attraction of FDI as the structural condition of Ireland's economic and industrialization strategies. Lemass's favorite dictum, "a rising tide lifts all boats," often attributed to John F. Kennedy, can be considered an ur-logic for economic common sense in Ireland. Throughout my research during the economic recovery of the 2010s, this phrase remained present in everyday life and public discourse as a metaphor for how the country's prosperity must be measured—still—along with the naturalized tides of global economic ebbs and flows.

These elemental metaphors of economic common sense help us to understand how pervasive these logics are in Irish governance. In their materialization through policy and infrastructure, cultural ideas and metaphors are *actually* connected to Irish state strategies and how they play out as tension and politics on the ground. Thus, what is important about tracing the popular history of Irish liberalization through this period is not what Lemass said, nor even the specific mechanisms that he encouraged and enacted while in power—nor even confronting the veracity of this "origin story" of Irish economic progress, which is obviously much more complicated and conflicted than the CliffsNotes traced above.[10] Rather, the origins of these economic logics pertain to a few specific and enduring resonances in the media infrastructural stories detailed throughout this book: (1) the metaphorical significance of Lemass's economic philosophy across Irish history and culture, to the degree that you would hear "a rising tide lifts all boats" from either a film studio CEO or a freelance worker, with varying degrees of assurance, hope, sarcasm, or exhaustion; (2) the establishment of Ardmore during his tenure as minister for industry and commerce, setting the stage for future offshored media productions in Ireland; and (3) the emplacement and facilitation of a system of political and economic liberalism, setting in motion a pervasive and prevailing economic common sense around the necessity of FDI for industrial development and economic survival. This "naturalized" mode of thinking manifests in everyday life and through a variety of governing institutions, rendered "sacrosanct" and untouchably important to Irish economic life (McCabe 2022).

With the forces behind the "rising tide" thus naturalized, its waves crested in the "economic miracle" of the Celtic Tiger and crashed spectacularly with the global crisis of neoliberalism in 2007–8. Thus, these commonsense naturalizations refer not only to the presence and role of FDI in shaping the political and economic landscape of the country. They serve as a visual heuristic for processes that subjugate culture, space, and labor to the turbulence of the global market and the epistemologies of value extraction that characterize what geographer David Harvey once referred to as the "financialization of everything" (2005). And it is in critical reference to this set of naturalizing logics that the basis of this book's approach to Ireland's media infrastructural politics is formed, addressing across its chapters the specific integrations of Ireland and its media infrastructures into the capitalist world system (chapter 1), the financialized logics of the "creative city" in recovery-era Dublin (chapter 2), the uneven and precarious spatial distribution of media industries across Ireland in relation to creative policy and tourism (chapter 3), and finally the climatic politics and circulations of big tech infrastructure in the form of data centers (chapter 4).

Media, Finance, and Irish Globalization

The key factors in the 2007–8 financial crisis in Ireland were speculative finance and property development, buttressed by the state's FDI policies with the help of favorable development zoning and tax incentives. When the property bubble burst, as will be detailed in chapter 1, the Irish state doubled down on neoliberal measures amid the turmoil to generate a recovery economy centered again on FDI—and deepening privatization—across most sectors. Largely due to "recommendations" imposed by the bailout conditions of the "troika"—the International Monetary Fund (IMF), European Commission, and European Central Bank, which implemented a loan and austerity program based on so-called fiscal discipline designed to pull Ireland out of debt—the country was compelled to slash public jobs and subsidies, privatize assets, and enter what was essentially a structural adjustment program by another name (Coulter 2015).

But how exactly do we get to a point that state-led neoliberal reforms across a diverse range of public-sector bodies—common across much of the so-called post-developmental world—become so inextricable from direct interventions by private-sector, export-driven industries like financial services and big tech? After the financial crisis, it became clear that any sector that could draw in additional investment and "jobs" to the starved economy would soon be reflected in policies, and this included the traditionally "public" media policy of the Irish Film Board/Screen Ireland, which extolled the agency's ability to facilitate foreign productions and enable partnerships with other sectors. While this might appear to be a textbook example of neoliberal privatization enacted through cultural policies, it is worth unpacking the core logics and material operations of this neoliberal common sense—and where exactly "culture" lives in this set of political economic strategies.

Neoliberalism, as a catchall term for these measures both in Ireland and imposed from abroad, before and after the crisis, became in the 2010s a straw person for arguments from the left, critiquing the values of highly marketized economies reliant on widespread privatization and multinational investment (among more aggressive measures). Ireland is no different, and "neoliberalism" as a term was thrown about in the aftermath of the financial crisis, referring to Ireland's "neoliberal crisis," the "invisible ideology" of neoliberalism, and other ascriptions of power to this economic and political system that had operated in Ireland for some time, but was only just entering the popular lexicon from the left and academia. While none of these analyses are incorrect, and I share and draw from them quite extensively, this book unravels the specific industrial threads that characterize contemporary Irish neoliberalism, focusing first on the financialization of the economy and its effects on spatial development and everyday life, before examining more closely its implications and deployment within media and technology industries.

Financialization, then, forms a backdrop for much of this book's analysis by *naming* and *identifying* exactly how certain strategies and orthodoxies traditionally ascribed to neoliberalism emerge in fields far beyond typical corporate actors and state spheres of regulation. This is in part due to the powerful presence of finance capital across the landscape of Ireland both pre– and post–financial crisis, and the naturalization of financial rationalities within the country's planning, society, and culture during austerity. At a basic level, financialization is, as literary scholar Max Haiven summarizes, "a term which refers to the increased power of the financial sector in the economy, in politics, in social life and in culture writ large. More expansively, the idea of financialization speaks to the way financial measurements, ideas, processes, techniques, metaphors, narratives, values and tropes migrate beyond the financial sector and transform other areas of society" (Haiven 2014, 1). Pertinent to the wider argument of this book, financialization is particularly urgent in the ways in which finance comes to organize spheres of life somehow thought to be far removed from financial considerations—not only property development, considered a typical and particularly heinous sphere of financial speculation in the aftermath of the financial crisis and the embroilment of subprime mortgages and other bad lending practices, but also media, technology, and fundamentally how people live, work, and consume and participate in culture.

But beyond more familiar industries like property, energy, utilities, or services, all of which have come under duress by privatization and speculation for several decades, less studied has been the financialization of media industries. Financialization can be read through most contemporary geographies of media, and this is not to mention the power of finance capital in Hollywood cinema (see DeWaard 2020). Even in more public systems like Ireland, government tax incentives attract foreign capital in the form of investment in media production, and mechanisms like special purpose vehicles (SPVs), short-term companies established for tax residency, were formed to make use of these exceptional financial mechanisms and funnel profits offshore. This then shaped spatial and industry practices in each territory, as infrastructure and facilities had to be built to accommodate these operations. In turn, property markets and deregulated planning dictated where global media capital would land in the form of these infrastructures. Financialization is not "immaterial," but rather describes a profoundly material process by which space and culture are transformed under the influence of finance capital.

As the story goes, "ripple" and "spillover" effects of these financialized practices create the infrastructures required to sustain the industry—local film companies, training and expertise, production and postproduction studios—and will give rise to a more stable environment within which a local industry will thrive. But these private infrastructures are supported and inflated by big capital, boxing out most upstart individual or company competition by absorbing and accumulating these transnational circulations. Within these logics, then, workers and many creatives are, ultimately, dependent on the precarious flows of media capital, left to compete for scraps from limited and unstable institutions, as the eventful waves of private finance promoted by public state policy are barely felt as a trickle for many across the country. As sociologist Verónica Gago argues through what she calls "neoliberalism from below," conditions of financialized *lack* of state investment and care, like the environment of post–financial crisis austerity that will be described presently, profoundly shapes how people relate not only to the state and the economy, but to one another. This means that politics must necessarily be theorized and practiced through an understanding of this multiscalar relationship between state power, global capital, and workers and communities on the ground (Gago 2017). Media infrastructures, as both strategic development priorities and crucial instruments of everyday services and experiences, offer a unique vantage point for analyzing how and where these relations materialize and operate within crisscrossing fields of power and tension.

The book uses this prevalence—or better, entanglement—of finance capital and its logics within these infrastructures of everyday life and work to tease out the specificities of Irish neoliberal development in relation to the postcrisis austerity economy. In a lineage from Ireland's early deregulated zoning mechanisms like the Shannon Free Zone (SFZ) and the International Financial Services Center (IFSC), Ireland's progressive "free zoning" from its establishment as a site for financial, media, and tech industry offshoring have contributed to the material conditions of Ireland at present and its geopolitical (and geo-economic) positioning within global trade flows. As I argue in various ways across the coming chapters, "logistical" governance, a logic that operates across state and corporate partnerships that optimizes the production of value across global supply chains (Cowen 2014), has become a transformative presence and actor across the Irish media landscape, creeping into state media and cultural policy as much as it operates openly through the facilitation of multinational, big tech circulatory infrastructures like Amazon Web Services (AWS).

When discussing the regulatory measures that states take in response to turbulent global conditions like financial crises, political economists like Mezzadra and Neilson articulate the ways in which the role of the state is pervasive, if altered and apparently withering, under shifting regimes of capital. In adjacent terms, the "national question" in Ireland has historically been primarily confronted by Irish cultural theorists in relation to the shifting and emergent social, cultural, economic, and political realities of globalized Ireland, especially in film and media studies (Barton 2004; Crosson 2003; McLoone 2000, 2008; Pettitt 2000; Rockett, Gibbons, and Hill 1987). While formative to how this book approaches the development of the Irish media industries, often secondary in these analyses is the constituent role of and interaction between the state and capital in the production of culture, much as critiques of national film and literary approaches have focused on the fixation with the "national" and its bias toward cultural identity and even sovereignty (see Kearney 1996). In this sense, it is not the focus on the "national" that seems a shortcoming of culturalist approaches, but rather the limited and limiting heuristic of national identity and sovereignty instead of *how a state thinks through territory* and *conjures certain conditions into being* (democratically or not). This involves seeing global capital as a political actor in dynamic relation to the state, as its role and power in governing populations, encapsulating "culture," and regulating capital has changed. This book understands the Irish state as not only an active but a *fundamental* historical participant in capitalist processes through operational modes and moments of (non)intervention. Only through such an approach does "the inability of the state to fully control or regulate the nexus of capital and politics" come into focus (Mezzadra and Neilson 2019, 52).

However, to understand the political power exerted by *capital* under emerging conditions, we have to remain attentive to how capital "crosses existing political forms" (Mezzadra and Neilson 2019, 52) like states and geopolitical relations in its multifaceted enactments across time and space. As Mezzadra and Neilson explain in more detail, "Recognizing the role of the state in the production of such conditions, as well as in the management of the challenges that become manifest when these conditions are destabilized, does not mean ignoring the wider array of government entities and actors involved in this process. However, the testing of government capacities within the crisis of the past decade has at once made states prominent again and revealed the limits of state actions and interventions as a means of restarting growth in a global situation characterized by the intermeshing and interdependence of economic systems" (49). These interdependencies affect spheres beyond typical areas of economic activity, such as heavy industry and finance, but intersect and affect the organization of sectors and ways of life thought to be shielded from certain kinds of intervention, traditionally conceived as the realm of "national culture" in relation to much longer and emplaced histories of practice and relation.

The long decade of 2007–20, which makes up the upper bounds of this book's temporal range of inquiry, provides a useful framework for understanding how the laboratory of crisis produces novel forms and arrangements of capitalist activity. While the preceding pages introduce how these processes in Ireland would not have been possible without the foundational experimental conditions created by almost a century of postcolonial (under)development (Bresnihan and Brodie 2025), "crisis" delineates a temporal period by which new and inventive ways of generating and extracting value through and from these geographical dynamics. Popular commentator Naomi Klein has described the ways in which capitalists cynically manufacture and exploit catastrophe as "disaster capitalism" (2007), a cyclical process by which the "free market" entrenches power and wealth into the pockets of an increasingly shielded few. This book extends this and similar hypotheses historically by suggesting both that Ireland has long been a kind of "laboratory" for particular forms of governance (see Deckard 2016) *and* that crisis, and financial crisis especially, drives terrible innovation in *fixing* flows of capital through evolving cycles of spatial practice. From a "spatial fix," as geographer David Harvey has famously theorized (2001), to a "creative fix," as chapter 2 refers to neoliberal creative policies, these market corrections are retroactively ascribed to economic recovery, though later revealed as empty veneers for capitalist exploitation. The decline of the "creative sector" as an exciting and viable career path in the aftermath of the long 2010s is a case in point (see Whiting, Barnett, and O'Connor 2022)—retrospectively, it becomes even clearer how the mechanisms of "recovery" analyzed especially from 2017 to 2019 were powered by the corrosion of public services and support across multiple sectors, largely in and through the cynical capture of cultural and media activity by and for capital.

But while globalization has often been seen as a problem to be dealt with in terms of culture, the largely turbulent shifts associated with globalization and neoliberalism have brought mixed blessings to Ireland. The Celtic Tiger intensified social and economic inequality within what was revealed in 2007–8 to be largely chimeric capital growth. Since the 1990s, more active participation in the world economic system has led to undeniably advanced levels of prosperity, during which Ireland has catapulted to one of the wealthiest nation-states by GDP per capita in the world, even despite years of downturn and stagnation following the financial crisis. It has also seen major progress on social and specifically gender-based repression, as the long shadow of the Catholic Church has begun to wane in Irish life, encapsulated by the passing of the Thirty-Fourth Amendment allowing same-sex marriage (2015) and the repeal of the Eighth Amendment prohibiting abortion (2018), both by popular referendum.

However, at the underside of these changes has remained structural inequality, housing shortages, the privatization of services, the racialized and carceral immigration and direct provision system, reckoning with the historical abuses of the Catholic Church, and ongoing and often conservative cultural negotiations about what it means to be "Irish" after globalization. Until recently, progressive values have not been represented through the electoral process, with the center-right parties Fine Gael and Fianna Fáil still dominating government through the crisis and recovery. Protest movements tended to be local and reactive in character, even as the countrywide "recovery" metrics looked more like "leprechaun economics" by the day (Regan and Brazys 2018). Regressive politics could still be found in both the margins and the mainstream, and many citizens who felt left out or abandoned by the state's Dublin-centric, FDI-driven, recovery-era prosperity engaged in the kind of affective and racist politics seen in the era of right-wing populisms in the rest of the EU, the United States, India, and elsewhere.[11] In the 2010s—and still now—the place of the occupied six counties in the north within the Irish state narrative remains unresolved. Republican party Sinn Féin's electoral successes in the north and the Republic in the early 2020s, and periodic talks of a border poll, demonstrate the degree to which the *national* question itself, like the uneven "European" modernity that has characterized the postcolonial Irish state, has never been fully contended with.

It is within the unfinished and constantly evolving cultural politics of Ireland that this book analyzes the role of media and its infrastructures in the post–financial crisis era. Political change (if not progression) has not prevented the Irish state from capitalizing on the essentialisms and com- modified versions of "Ireland" as a product that has been the bugbear of many cultural critics, especially in film and media studies (Barton 2004; Pettitt 2000), demonstrating the ways in which the cultural bias of the "national" has often acted as a funnel for commodification. Ireland has long been "branded" for international audiences, and romanticized Irish imagery and flippant "paddywhackery" remain persistent and powerful within consumer culture in Ireland and many parts of the world (James 2014; McGovern 2002; Mulhall 2013).[12] Beyond tourism, however, the Irish state also extends this cultural "soft power" (Nye 1990) into how it operates and attracts foreign investment in general, branding the culture, state, citizens, and space of Ireland as ideal resources for the extraction of value via industry-adjacent groups and semi-state organizations like the National Asset Management Agency (NAMA), the Industrial Development Authority (IDA), Screen Ireland, Culture Ireland, Enterprise Ireland, Host in Ireland, among many others. I argue that these organizations, whatever their remit, employ and instrumentalize these cultural values toward a regime of FDI-driven profit.

This book thus contends that, by experimenting with and intensifying these earlier forms of commodification and development, the post–financial crisis Irish state posed these facets of Irish life and its labor force as "re- sources," employing regressive and essentializing cultural logics across long-standing political fault lines. Joining with the private sector, then, the state peddled an ongoing series of naturalizations with regard to so-called "uneven development" across Irish space, culture, and labor (chapter 1), whether by unquestioning commitment to financial logics in urban spatial development (chapter 2), its treatment of Irish labor as a "resource" for in- ternational media production (chapter 3), or the artificial generation of a "business climate" suitable to the extractive environmental infrastructures of the tech industry (chapter 4). Throughout the language, policy, and planning materials deployed to support such projects, these conditions are posed as *natural* elements of everyday existence in post–financial crisis Ireland. In a way that is instructive to other contexts, this book critically assesses the persistent and ongoing naturalization of FDI-driven logics and cultures of neoliberalism across the country's public social, political, and environmental discourses.

Entanglement at the Edges

Naturalization, after all, is not *only* something that can be revealed through critical analysis, lifting the veil of ideology to reveal the truth underneath. The representation of anything as a natural set of conditions, especially within *state* strategies, has very real material effects and social histories. Especially when the problems of an economic system are revealed—for example, during the aftermath of the 2007–8 financial crisis—"extractive rationalities need to be naturalized or normalized" (Couldry and Mejias 2019) to resume function without delays, objections, and oversight. In Ireland and elsewhere, especially during austerity and privatization, it is a tried-and-true strategy by state and corporate organizations to rhetorically associate their actions with benign or affectively appealing natural processes.

Take, for example, a 2016 Deloitte pamphlet entitled "Waves of Disruption: The Future of Ireland's Financial Services Sector" (Dalton and Marmion 2016). Throughout this report, the authors determine that it is "environmental" aspects of Ireland's business world—its regulatory "climate"—that must be controlled and managed to respond to the "disruption" that occurs in global tech and finance markets. These disruptions are created by a series of factors, according to the report, including the regulatory changes brought on by the 2007–8 financial crisis and the resulting market changes, like digitization and "consumer empowerment." The *climate* of a place can be controlled through savvy business (and regulatory manipulation), whereas consumer activity cannot, betraying how companies conceive of market control as a method to rein in and predict the activity of unruly citizen-consumers. This pamphlet, like other such pieces of "gray media" and otherwise "boring" corporate communication studied for this book (Ballestero 2019; Bowker and Star 1999; Opaque Media 2017; Star 1999), provides insight into the cultures that sustain and produce the above-mentioned naturalization, providing routes for ongoing capital circulation and accumulation.

Wild Tides, in transparent conversation with gray media like "Waves of Disruption," analyzes where and how media centers within naturalized "cultures of circulation" characterizing contemporary capitalism (Lee and LiPuma 2002). Circulation functions through the operational and infrastructural conduits built (physically or legally) to facilitate it, the "ambient environment of everyday life" (Larkin 2013, 328), which appears natural but is manufactured by policy, planning, construction, and capital. The book locates several ways in which this functions in post–financial crisis Ireland—whether in chapter 1, which demonstrates the historical and spatial development of Ireland's FDI-driven media economy; in chapter 2, which uncovers how financial circulation operates through the built environment of the "creative city"; in chapter 3, which discusses how global media supply chains capitalize on regional competition and the FDI-hungry logics of film and media policy; or in chapter 4, on how server racks and cable networks of digital economies act as conduits and grounding points for technological, logistical, and financial supply chains.

Turbulence, the choppy waves and shifting tides of the world and its economies suggested even by the Deloitte industrial literature, must always be managed for capital to operate and extract value through instability. Finance, infrastructure, and logistics are ways in which to exert some manner of control over a dispersed and often unmanageable assemblage of global factors across spaces and contexts, and their logics have proved robust for states and corporations attempting to govern these flows to the best of their abilities. However, in the process, their entanglements have become naturalized across both top-down discourses and in many other social and cultural contexts. Media industries and their infrastructures are increasingly at the forefront of how the future of global economies will function, especially through the tech industries and their reorganization of economies across the world via digital software and hardware. The horizontal integration of media and technology, across diverse supply chains and contexts of circulation, demonstrate the degree to which media conglomeration and technological development will organize and disorganize across a diversity of global sites, institutional and infrastructural formations aggregating and disaggregating along with the turbulence of global systems. These private operations become a strategy for *managing* turbulence, coming to visualize and spatialize the experience of turbulence itself.

This tumult and its governance begin the process of materialization in gray media. In the "boring" plans and images for these infrastructures, exchanged across desks and emails and discussed in boardrooms and offices until eventually becoming plastered on the construction hoardings and billboards for new projects, visual media of planning and development demonstrate *financialized* and *logistical* ways of seeing and visualizing the future, which imaginatively and physically foreclose alternative and existing lives and visions for environments. They perform several different kinds of erasure, all at once: the erasure of existing communities and ways of being in these spaces; the augmentation of spatial histories and ruins with capitalized futures; and finally, the removal of the labor that goes into these sorts of plans, whether in real or virtual space. If finance and logistics, as dominant forms of circulation across the global landscape, are shaping the networks within which nation-states and regional units exist, control the trade of goods, and build economic power, we need to use all tools at our disposal to understand how these forms of circulation are visualized, come into being, operate, and collide with and through the state and diverse communities on the ground.

Researching infrastructure, as anthropologist Susan Leigh Star has influentially theorized (1999), thus means devoting ample time to these "boring" chores of analyzing not just technical materials. It also means chewing through arcane, jargony, and niche government documents, policy briefs, community consultations, reports, corporate strategies, court cases, promotional materials, and the gray media that characterize the everyday work of politics and business, paying close attention to the power hierarchies expressed within, visualized, and reproduced by their future-driven language, plans, and images (figure I.1). Media scholar Rafico Ruiz demonstrates that such documents, in his case archival materials of a Canadian settler colony, are themselves a kind of "media infrastructure" in the ways that they organize and act to institutionalize practices, physical spaces, and ways of life (2021). While much of this book draws insights from sited fieldwork, these visits were supported and enriched by months spent combing industry reports, plans and designs, policy documents, newspaper articles, and state and corporate organizations' websites to grasp how promotional and strategy logics were being operationalized through on-the-ground media infrastructural conditions. Contained within these documents, then, are the futures that the state and capital attempt to implement. They imagine worlds that appear beneficial to public goods and needs. However, as visual representations of financialized life and culture, they are rife with unseen and inconvenient spatial inequalities and violence. They show us the logics that they naturalize, if we know how to look at them.

Financialization and the logistical organization of the global economy mean that such plans are used to "grease the wheels" for future operations, mobilized by the state and capital.[13] However, as "plans," they are inevitably disrupted by the messiness of life and experience. Top-down planning is experienced very differently on the ground, and such conditions enrich or disrupt capital's operations. In the process of writing this book, I performed site-specific fieldwork and interviews between 2017 and 2019. I visited job sites, community organizations, individual workers, government offices,

I.1 Digitally rendered plan for AWS data center in Drogheda. Data Centre Dynamics. https://www.datacenterdynamics.com/en/news/environmental -group-blocks-amazons-second-400m-data-center-in-drogheda-ireland/.

and corporate infrastructures. The book in its current iteration thus balances the speculative and operational gaze of transnational capital with sited engagements at points where these materially are felt and enacted. But the translation from field to manuscript is never a smooth process, and thus this book reflects throughout the methodological messiness of on-the-ground research and the often-cumbersome task of fitting real-life encounters into the conceptual molds required by an academic monograph.

Fieldwork, as I reflect upon throughout this book, is a useful method for muckraking and mapping out stakeholders across localized and transnational environments. It cannot alone answer for all the nuances of subjectivity, the messiness of politics, nor the precise ways in which value is extracted from emplaced environments. Echoing media scholars Lisa Parks, Lindsay Palmer, and Daniel Grinberg's reflections on sited media studies research, fieldwork "coincides with an interest in understanding how material conditions, location, *difference*, and *power hierarchies* function as part of media cultures" (2017). Difference as a mobilizing structure for supply chains and the power they materially operate through—financial, infrastructural, logistical—are primary issues for this book. But the pictures that scholars paint are necessarily filtered through theoretical and methodological tools at our disposal, documents we have time to study and sites we can feasibly visit, and our very positions as researchers legitimized (and limited) by often extractive institutions, risking the reproduction of such hierarchies in our endeavors. This leaves the task of weaving together theories, documents, discourses, and lived experiences as delicately and ethically as we can. There is always a politics to how these things are brought together, what is represented, and what is left dangling outside of our study.

In this way, what this book does is tactically work across scales by leaving space for unexpected and recalcitrant infrastructural politics arising from unique experiences of their development, operation, and failure. The spectacular presence and future dreams of media infrastructure, what Larkin acknowledges is a kind of infrastructural "sublime" when dealing with large-scale projects (2008), necessitate a sustained and measured engagement with diverse and unexpected experiences of media infrastructure. Despite the mundaneness of their plans and visualizations, the spectacle of development—whether through the extended and often antagonistic planning process, news reporting, or large-scale construction—directs "eventful" attention away from the everyday precarity and abandonment of people conceived as or left outside of them (see Povinelli 2011).

Rather than seeing the large- and the small-scale as separate objects of study, however, throughout this book I identify multiscalar tensions and entanglements occurring across sites, communities, and institutions, which are often not as antagonistic as you would expect. Individuals and communities, wittingly or unwittingly, are frequently powerful tools and advocates for capitalist systems. Anthropologist Aihwa Ong addresses the often-ambivalent forces affecting the cultural politics of economic systems, choosing "to examine the everyday effects of transnationality in terms of the tensions between capital and state power because *there is no other field of force for understanding the logics of cultural change*" (2000, 23, my emphasis). The bluntness of her advocacy for this approach remains inspiring: When it comes down to it, the state and capital—and their logics of value and governance—have power over our lives in more ways than we care to admit, even as we live through different vectors of precarity or stability, whether in terms of jobs, racialization, abandonment, or environmental collapse. We relate to their infrastructures every day, and our behaviors and politics are influenced by these relations. When these structures are transnational in scope, and increasingly formally discrete from public institutions and forms of care, how exactly do we then form a politics around them, whether in acquiescence or struggle?

But just as state and corporate power are black-boxed and inaccessible, local politics are equally recalcitrant and hard to grasp, making it especially difficult to follow strands between these levels of operation. In Ireland as elsewhere, local culture and politics are crucial to understanding how media and tech companies generate consent or dissent for their territorial projects. Communities do not think the same thing collectively; they consent or dissent in varying and antagonistic ways, compelled or choosing to relate to the state and capital in whatever way seems to serve the interests of their needs and beliefs at any given time. Thus, both *scale* and *place* are central to research into media infrastructure projects. As environmental humanist Rob Nixon argues, local environmentalism and advocacy tend to have a hard time "scaling up" from local politics to more global or planetary concerns (2011). For example, local anti–wind farm groups in Ireland have trouble translating their localized concerns—biodiversity destruction, lack of consultation, noise pollution, unequal distribution of impacts and benefits— to a planetary or even national scale, which tends to privilege the urgency and immensity of global climate change over these small-scale concerns.

At the opposite end of the spectrum, powerful corporations are enormously scalable (see Vonderau 2019), crunching place into their global metrics, incorporating these local frictions into their very way of generating and extracting value. However, as anthropologist Anna Tsing argues, studies of global capital have often failed to explain the necessary and structural diversity of transnational economic systems, and we must better account for the "bigness" of this operation across different scales, an "imaginative project" for both those who study and those who enact the power of global capitalism (2009, 153). Much of this book focuses on the machinations and places where this big capital "hits the ground" (Mezzadra and Neilson 2019) in Ireland—for example, in Athenry, County Galway, where a proposed €850 million Apple data center was supported by a popular movement of thousands called "Athenry for Apple," but was stopped by a small minority of objectors on environmental grounds; or in west Kerry, where Disney filmed portions of *Star Wars: The Force Awakens* (J. J. Abrams, 2015) and *The Last Jedi* (Rian Johnson, 2017), shooting on UNESCO World Heritage Site Skellig Michael before international pressure forced the IFB to revoke permission, while still allowing them to film in other rural locations across west Kerry. Local infrastructural conditions are overhauled by such moments of collision, whether in terms of the social fabric behind infrastructure (like civil society, in the case of Athenry) or physical infrastructures themselves (the film crew for *Star Wars* had to build modular, temporary access roads across muddy fields to get to inaccessible coastal sites in west Kerry [figure I.2]).

But as Mezzadra and Neilson articulate, it is not enough to understand these places as simply where capital "hits the ground," even if this is a useful heuristic. Rather, we must see them as sites through which to "move from the subterranean to the surface level and back again to show how the systemic edge is always caught in a dense fabric of frictions, conflicts, and resistances" (2019, 138). In doing so, we can understand the entanglements on the ground that drive such projects and then also disrupt, challenge, or support them. If, as media scholar Ravi Sundaram argues, "infrastructures are at the center of media circulation by way of entangling people, objects, knowledges, and technologies" (2015, S299), then these are not just incursions but immediate *entanglements* with social, political, economic, cultural, and environmental conditions.

Workers and communities operating within these entanglements, while actors within infrastructural assemblages, are often thrust together by these very same impersonal arrangements. In one day I spent doing interviews in Limerick city, I was shepherded around to various arts spaces, coffee shops, bars, and organizations, as I followed the tangled threads of local actors within the media and arts scene, all tenuously interconnected

I.2 Infrastructure for *Star Wars* film set on Ceann Sibeál, Ballyferriter, County Kerry. Irish Examiner. https://www.irishexaminer.com/sport/golf/arid-20464660.html.

by the infrastructural investments (or lack thereof) in Limerick's "creative industries" that I was trying to unravel. In a day spent in Athenry, I was similarly directed from storefront to storefront, and even to someone's front door in the town center, pointed to new and various people and organizations involved in the town's civil society with each additional meeting. Years later, I still occasionally received Facebook messages from Athenry for Apple members I met or was put in touch with that day, who asked me to advocate for future data center projects in Athenry via my social networks. When meeting with a data center developer in Newtownmountkennedy in County Wicklow, our conversation was interrupted by a phone call from a former state official with whom he was working. The developer handed me the phone, and the former official conspiratorially avowed to me that the country was run by a cabal of powerful, wealthy people who also controlled RTÉ (the national broadcaster) and were insulated from democratic accountability by controlling public opinion through the national media. Throughout my research, word of mouth, community networks, and general presence have been inordinately more valuable and generative for thinking than expertise—even just for odd stories like the one above. Dozens of similar encounters never made it into these pages, but nonetheless inform a robust background of my approach and understanding. There are always kernels of truth and insight to be gleaned from such accidents and entangled experiences of place, capital, and community.

Theorizing the "Wild"

Film and media studies' methodological and conceptual training has more often favored historical, discourse, and textual analysis than these sorts of on-the-ground engagements with sites, communities, and technologies. In Ireland, when studying cultural forms, this has frequently involved the use of media texts as an insight into cultural change and social organization, especially since the growth of a publicly supported film industry from the 1970s and 1980s and the national transformations of the Celtic Tiger (see Barton 2015; O'Connell 2010). Building on this, however, this book is inspired by Star's theorization of the diverse practices required to perform ethnographies of "the imbrication of infrastructure and human organization" (1999, 379). I read this imbrication through projects of spatial development, culture, industrial activity, and state strategy around media as a social and technological formation rather than traditional media texts themselves.

That said, trained as a film and media scholar, I am no ethnographer.[14] Parks, Palmer, and Grinberg's insights on media studies fieldwork are useful to my methodological application: "Our fieldwork does not aspire to or adhere to all of the central tenets of a classically defined ethnography: immersion in spaces, mastery of languages, establishment of cultural competency, longitudinal study, participant observation, and 'thick description.' As media scholars we are primarily interested in understanding how diverse communities in the world think about, organize, and use media technologies to support their interests" (2017). When it comes to media politics, there is always more to the picture. Panoramic views can sometimes foreclose the real and productive messiness of life and labor under financial and supply chain capitalism. Unfortunately, I could not speak Irish to people in the Gaeltachts (Irish-speaking regions) in west Kerry and Galway, nor would I ever purport to understand the granularities of context in a way that only someone who has lived within that community for years or decades possibly could. Neither do I know in the same way as lifelong or permanent residents the rhythms, the relations, the everyday feeling of a life lived in Dublin, Limerick, or Athenry.[15] This must necessarily be addressed as a shortcoming for the sorts of work this book does.

Media infrastructures, however, offer sites through which to begin an analysis of media's material politics—a data center, for example, or a film set, or an "undeveloped" plot of land under the planning gaze for the creative city. This targeted approach thus gives space to the unexpected, and what my own perspective can bring to the study of Ireland's post–financial crisis media landscape. Where are some of the places that financialization has not been as widely theorized, for example, the rural media industries in which I spend much of the book? What does it look like, and who does it affect? What are the weird formations of financialized media in Ireland that only reveal themselves to an outsider's perspective? And how are the logistical forms of governance identified throughout this book enrolling these differential media landscapes into processes of capital accumulation? These are questions that apply and look different whether you are looking at a graffitied wall in central Dublin, or a forest for sale in rural Galway.

In this way, the book also responds to what Jack Halberstam calls the "metronormativity" of humanities research (2005), which reflects an urban bias of media studies associated with core issues characterizing globalization and connectivity: for example, speed, modernization, and emergence. But as Parks, Palmer, and Grinberg also note, media infrastructures are often *rural* spatial technologies designed to *support* processes of development and urbanization *elsewhere*—for example, energy infrastructures supporting urban technologies like data centers and smart sensors. These forms of infrastructural mediation of spatial and political relationships provide powerful insight into the durable structural imbalances, exclusions, and extractions of contemporary capitalist (and colonial) formations (see Brodie and Barney 2026; Ruiz 2021). If cultural studies teaches, in the words of Stuart Hall, the basic point that "conditions of existence [are] cultural, political and economic" (2007, 156), and geography, according to Ruth Wilson Gilmore, examines "why things happen where they do" (2020), combining these approaches helps us understand interconnected histories and industrial formations of media and economic development *through* Ireland and its emplaced iterations.

Where infrastructure "hits the ground," unruly and conceptually disruptive worlds reveal themselves. This is where we must center the unexpected, both challenging the promised systematicity of capital while recognizing the multiple, messy avenues and directionalities that arise out of infrastructural encounters. Strange and uncanny entanglements occur in meetings with local communities and their embedded political lives and affects. The following accounts of these instances are sketchy, of course, but necessarily so, because that is how experience is, and throughout the process of writing this book I have tried to preserve the character of these interactions.

In her study of energy politics in the Orkney Islands in northern Scotland, Laura Watts deals with similar assemblages of people and things in looking at the speculative financial gazes that are focused on Orkney due to its environmental resources (a test bed for tidal power), with tech investors, European and UK politicians, and local communities operating across the various scales of Orkney's social, political, and environmental life. While sharing an office waiting room with a Silicon Valley entrepreneur looking to invest in Orkney-generated energy, she describes events with the following qualification:

> this is not precisely how it happened, not precisely how it was said, because all empirical research can only record so much. Stuff happens off camera, the pen can only move so fast, you can only sit in one chair, not all the chairs in all the rooms. This is good, honest objectivity because it has good, honest limits. . . . What you read has a *partial perspective*, as theorist of science and technology Donna Haraway long ago named it. Its knowledge is situated, embodied, in relation to others,

because the knowing that I am after, what I want to tell you about and work with, has no laboratory walls nor well-defined variables. It is a communal endeavor, and the social world is always a right *mess*, to paraphrase sociologist John Law. . . . This story honors that rich tangle by looking at it with both eyes, with a *split vision*, and finding different strands to follow. (2018, 5–6)

To lock these encounters into place, to gift them to the systematicity of capital would be to reproduce the governing impulse to *control* and *manage*. Everyday life is wiggly, and infrastructures and governance struggle to keep up, to make everyday worlds knowable, legible, and manageable. Global turbulence may be the norm and appear natural in and of itself, but the "nature" of everyday life is uncontainable and unmanageable. Infrastructures, by their definition, *circulate* things through environments; governance *manages* how this happens. But people do not live and move according to fixed structures. Rather, in Ireland, as with many other places with their own complex and layered infrastructural and political histories, strange encounters and unexpected connections abound, and this should excite rather than limit our approaches.

Even if this "wild" world at the edges of the frame remains constitutively offscreen of written research,[16] we should remain skeptical of attempts to manage it, especially within our own practices—and I am being self-reflective here, not least due to the title of this book. "Wildness" exists within a colonial lineage of frontier designations, territorializing and applying the "unruliness" of bodies and environments for violent incursion by capital (see Cram 2021; Yusoff 2018). Halberstam and Tavia Nyong'o recognize that the "wild" refers to "a colonial division of the world into the modernizing and the extractive zones" (2018, 455), reflecting the dynamics of spatial development that subjugate people and places to the whims of global capital. Ireland, as a postcolony,[17] is still often seen through colonial imaginations and frontier logics of its rural, "wild" backwardness, a place of precolonial heritage and colonial extraction, and this is oftentimes peddled by the Irish state itself as a development strategy. *Wild Tides* as a conceptual gesture associates the turbulence of global markets with the cultural and environmental politics of Ireland and climate change, doing so to comment on the persistence of colonial optics in contemporary Irish infrastructural development politics.

The "wild" is prevalent, for example, in the popular "Wild Atlantic Way" tourist campaign, which has branded the west of Ireland as a differentially tamed, authentic landscape and culture to *experience* by driving up and down its coastal road network, supporting local businesses and consuming culture along the way. It has been successful in drawing international tourists and business to the Atlantic coast in the post–financial crisis era, and was a relatively popular initiative during the period of my research. But it also, in clearly nonreflexive ways, by treating the Atlantic coastline and its cultures as "resources" (Government of Ireland 2018b, 103), commodifies the people and landscape via regressive—and colonial—visions of rurality and authenticity. Theorists like Doreen Massey (2005) have argued that "uneven development" conceptualized by scholars of globalization like David Harvey (2005) and Neil Smith (2008) does not only mean that underdeveloped regions are disempowered. Across Ireland and elsewhere, there are differential experiences of economic and spatial development and responses to it, filtered through and coproduced by complex and emplaced social, cultural, and environmental histories. To these scholars, the so-called "waiting rooms of history" (Chakrabarty 2000) imagined by theories of uneven development discount the already-global forms circulating through such environments (see Woods 2007), the turbulence and crises that are felt in differential ways across urban and rural spaces, and the ingenuity of people living so-called underdevelopment and abandonment. The same state forces that have traditionally left rural places behind in development, promising a prosperity that never arrives, now expect them to fend for themselves under the cruel care of a turbulent global market (see Berlant 2011; Gago 2017; Povinelli 2011). Austerity breeds unexpected relations and endurances in the face of global capital, whether through popular politics and economies, or acts of everyday care and getting by (see chapter 3).

However, there remains a seductiveness in the "wild," especially its aesthetics and anarchic suggestion of difference truly outside of colonial and capitalist regimes, unable to be "captured" by the rationalities and normativities of modernity. Halberstam has written powerfully on this subject (2020), especially in its suggestion of queer lines of flight and ways of life. His treatise with Nyong'o on wildness as a generative political force, articulated through a series of articles in their 2018 special issue of *South Atlantic Quarterly*, proves that under the right circumstances, the "wild" remains something to be sought, at least under (and outside of) the punishing regimes of colonialism and capitalism.

But rather than building on Halberstam and Nyong'o's aesthetic critique, which expands the potentiality of the "wild" and challenges its relationship with modernity, *Wild Tides* sits with the term's ambivalence, and how processes of representation through the media of policy, planning, and infrastructure interact with histories and presents of territorial boundary-making and breaking by powerful forces of global capital and its partners in the nation-state. Ireland's "wilds," like those elsewhere, have always been sites of strategic territorialization, whether for resource extraction, conservation, agriculture, tourism, or other forms of development. The difference in Ireland is that this powerful legacy of the "wild" permeates even its urbanizing and industrializing development paradigms, a kind of neoliberal "wild west" where deregulation and culture collide to create coalescing circumstances for turbulent growth. Experimental relations emerge and take form through Ireland's spaces and infrastructures—in the timeline of this book, especially during the global financial crisis and subsequent "recovery." *Wild Tides* thus tells the story of systems by which the public and private institutions worked together to use culture, media, and technology to capture, control, and instrumentalize both the turbulence of global capital and the unruliness of culture and environment, foreclosing the spatial cultures, livelihoods, and work that could not be made profitable. In doing so, this book shows that Ireland's "wildness," far from a site of mere colonial territorialization or a vibrant and unruly subjectivity, is constitutive of the neoliberal character of spatial and economic development itself. The maintenance, management, and harnessing of "wild" circulations, via various governmental, cultural, and business-driven means of naturalization and instrumentalization, is crucial to the creation and extraction of value from Irish space.

In this sense, it is not only global "turbulence" being managed—this turbulence generates conditions of possibility. Geographers Charmaine Chua, Martin Danyluk, Deborah Cowen, and Laleh Khalili draw a generative dialectic of "turbulence" and "liveliness" under global logistical systems (2018), where economic and political turbulence is met with the productive vibrancy of local communities and histories. As anthropologist Cymene Howe also argues in her study of wind energy ecologies in Oaxaca, Mexico, focusing on how elemental circulations like wind, which breathes life into renewable economies, "assembles" entangled threads differently and helps "to look for a new, turbulent prototype" (2019, 3) of politics. In this case, "turbulence," while disruptive to human economic endeavor, is also a generative, elemental force through which to understand alternative circulations of culture, power, and capital.

Turbulence, like wildness, must be *harnessed, managed, controlled* for it to be useful for capital, whether through the state, financial, or logistical networks. By conceptually enfolding turbulence and wildness, then, the latter is perhaps alone a twofold framework through which to understand both the system, state, financial, and logistical means of control and management, as well as what Mezzadra and Neilson call the "systemic edge" (Mezzadra and Neilson 2019)—somewhere at the margins of the supposed frame where we can acknowledge the non-totality of the system itself, while appreciating that these same systems leak and seep into everyday life (see Anand 2019; Liang 2005). Whether a highly specific land dispute in Kerry between two rival landowners about where Luke Skywalker's cliff-edge death scene was shot and who was allowed to charge for access to it; or the drunken nights at cultural events that fostered follow-ups with new field contacts in Dublin and Galway; or the Coillte official who asked me, an early PhD student in Québec, if I knew anyone who would be interested in a data center development opportunity; many experiences opened apertures to new and strange understandings and insights, while also presenting persistent dilemmas as to what to keep out of the final picture. This book takes us through a number of these worlds at the so-called "systemic edge" in Ireland, where one is compelled to realize that apparently stable media geographies are constantly formed and reformed by interactions among the state, capital, people, communities, and environments, whether disrupting or entrenching existing hegemonies.

When relating to the "worlds" you enter as a researcher, amid the crisis-laden and turbulent environments of global capitalism, we also must keep in mind that there are many different "arts of living on a damaged planet" (Tsing 2015). As Tsing notices, many across the world, in the Global North and Global South, are living "life without the promise of stability" (2). In Athenry, I encountered residents who craved the stability of a bygone era (industrial capitalism, which may have never fully arrived in most of rural Ireland) through a tech services–focused future. In opposition to the turbulent "wildness" represented by the financialized global economy and the livelihoods that it offers, the promise of jobs and infrastructure through these systems was seductive and all-encompassing, even if empty at the center. If modernity and modernization were designed to "fill the world . . . with jobs," "such jobs are now quite rare; most people depend on much more irregular livelihoods. The irony of our times, then, is that everyone depends on capitalism but almost no one has what we used to call a 'regular job'" (Tsing 2015, 3). Fieldwork among media professionals revealed a widespread acceptance of such unstable forms of work, which characterize the sector. However, although many felt browbeaten, others embraced the competition, passion, and flexibility that drive value production in the media industries (see Caldwell 2023). But while such arts of living through precarity sometimes move toward emancipation, and critical theorists tend to disproportionately focus on these more hopeful moments and practices, my fieldwork revealed that the politics are far more ambivalent. They short-circuit easy left–right distinctions and solutions, especially when it comes to labor, communities, and land relations. There is only so much a leftist research practice, defined by Mezzadra as "an attempt to localize, within a specific situation, the points around which practices of organizing and struggle can match" (2013, 310), can do in the face of the flooding of influence of commonsense logics of global capital enacted and managed through the state. Such promises of better lives through capitalist development are seductive. Sometimes you just have to work through what you see.

But this disparity is generative—unpredictability and contingency should be central to our methodologies, even if *stability* remains a chimeric fascination and goal of the institutions and communities we study. Research into these formations requires a simultaneous understanding of the ongoing allure of more stable and prosperous ways of life, which often take on unexpected and contradictory political formations, and the impossibility of achieving more just ones under the turbulent conditions of global capitalism.

Unweaving the Tethers

The threads holding this book together—to weather its own turbulence—come from structured fieldwork as much as the unexpected and even forgotten encounters that pervaded it. They are woven with structured and unstructured discourses and conversations with interlocutors and collaborators about media infrastructure, the state, capital, and the environment.

Infrastructures are hard to pin down, because they are as much about relations as they are about physical spaces. As Marxist sociologist Henri Lefebvre famously argued, "spaces" consist of social and political relations that *produce* our experiences of them (1991). Coursing through the environments and politics of contemporary Ireland are much longer histories, as well as the forms of subterranean power that continue to feed infrastructures built and sustained on extraction. *Wild Tides* may often give more space to power structures than other books that profess a similar political—and methodological—agenda. But looking at financialization across Ireland's media infrastructural formations, the built, imagined, and lived architectures of these spaces become clearer, as the structures of dependency and power within the world system emerge more clearly in and through Ireland's industrial landscapes. As each chapter analyzes, these tidal and climatic forces—both global economic and planetary—enact power and agency *over* given environments, continuing extractive colonial relationships in ways that appear natural and even beneficial to people that both do and do not directly benefit from them.

Chapter 1, "Turbulent Waters: Media Infrastructure and the World Economy," picks up where we left off with Seán Lemass and the early liberalizing state, expanding our gaze to understand the place of Ireland in the modern world system and the development of its media industrial formations through geopolitics and territory. Tracing the histories, infrastructures, and theoretical threads that remain essential for understanding contemporary Irish media policies, economies, and spaces, the chapter picks out a series of interconnected conditions and phenomena shaping the country before and after the financial crisis—neoliberal financialization, the media and creative industries, technological infrastructure, and logistics and extraction. The chapter thus lays the groundwork for understanding the Irish media environment's entanglements with global financial and logistical systems. Beginning by connecting early experiments with liberalization in the country to the deregulated zoning mechanisms like the SFZ and IFSC that carried through the recovery period, the chapter reveals the political implications of financialization through the 1980s and 1990s for the post–financial crisis shifts identified throughout the book. I then unravel the logics of the so-called "creative industries" in relation to post–financial crisis contestations around media and culture, and how they extend physically into the built environment through "media infrastructure" like film studios, broadband networks, and data centers. At this point, I contend that the naturalized, "ambient" circulations facilitated by infrastructure represent a point at which to identify changes in governance where the public and the private collapse within infrastructure and its management, demonstrating the prevalence of "logistical" forms of rule in new technological systems, particularly within data centers and other technological infrastructures.

Chapter 2, "Ghostly Currents and Creative Erosion," identifies the financialized logics at work within the built environment of recovery-era Dublin through direct encounters with the creative and media industries and the projects designed to support them. The Irish financial crisis was deeply tied to property markets, and it left swaths of half-built and unfinished housing and mixed-use developments across the country. These ruins of finance capital became a major cultural touchstone and public point of negotiation around feelings of guilt and hopelessness post-crisis. Within this aftermath, the Irish state established a for-profit semi-state organization called the National Asset Management Agency (NAMA) to absorb millions of euros in toxic real-estate assets across the country to make them profitable once more, whether by selling them, developing them themselves, or some combination of the two, usually in partnership with private and multinational corporations and investors. A major element of this strategy was the continued development of the Dublin Docklands, a former port area that houses the IFSC, an offshore financial services hub with a formerly exceptional 10 percent corporate tax rate, marked by several Strategic Development Zones (SDZs), or areas demarcated for private development with deregulated planning rules. The Docklands through the crisis and recovery transformed into a "creative city" and was home to the offices of many of the major tech multinationals located in Ireland, including Facebook, Google, and Airbnb. The technocratic asset management strategy represented by NAMA, informed by recommendations from the "troika" and neoliberal commonsense solutions to crisis, foreclosed alternative uses of space in the capital and imaginations of different futures to be built out of the crisis. Navigating the ways in which creative industries–led growth permeates the city, particularly through the visual media of planning and development and the capture of creative labor through spatial planning and policy, I unpack the place of media within the logics of neoliberal recovery and deep imbrications with space and culture by emphasizing the invisible circulations of finance and the spectacular experience of the "creative city." In discussing the ways in which affect, labor, and contingencies are captured and contained within Dublin's post-crisis built environment, the chapter shows how the spatial violence of finance capital circulates through living and ruined infrastructures across and through the creative city.

Building on the intersection of space and policy in chapter 2, the third chapter, "Waves of Austerity: Film Policy and the Infrastructural Geographies of Media Labor," articulates the ways in which media policy is *infrastructural* to the circulation of global capital through the country and its modes of extracting value from space and labor. Looking at how the policy of the Irish Film Board (IFB)/Screen Ireland (SI) post–financial crisis has become increasingly profit-oriented as the public remit of "cultural goods" has been sidelined, this chapter unravels the role of the state in managing national industries in the face of media globalization and austerity, articulating the relationship between state media industrial policy, infrastructure, and labor.

While many approaches to global media have focused on deterritorialization and the declining role of nation-states within transnational formations, I argue that the state is crucial to how private capital comes to colonize the logics of public good in given places, in this case through culture. However, while this complicates ideas of media sovereignty, my primary focus turns from policy to labor—that is, to how workers across the country feel and relate to top-down media policies—utilizing insights gathered from interviews and informal conversations with a range of media workers and professionals between 2017 and 2019. On a global level, the Section 481 tax incentive has brought in large-scale media productions from all over the world, and the state puts in place infrastructural conditions for them to operate smoothly. But most workers do not directly benefit, despite arguments about ripple and spillover effects. The landscape of media work in Ireland is characterized by precarity and contract-to-contract work, within which most media workers spend most of their time gig to gig rather than in the stable employment promised by FDI-driven media policy. The Irish state, within these policies, treats workers as a resource for value extraction, much as they tend to treat space as agnostic to whichever company is currently paying to use it. This positioning of workers as resources through ideas of "talent" and "skills" contributes to a naturalization of precarity at the whims of transnational investment and the supply chain organization of contemporary media economies.

Bringing together the creative industries, media circulation, and finally the place of big tech corporations within these ecologies, chapter 4, "Storm Clouds: Technology Industries and the Climate of Crisis," builds on debates forwarded in the first three chapters and identifies the convergence of a variety of state and corporate logics within a single, private infrastructure that has been at the forefront of planning and economic development since the financial crisis: data centers. The tidal forces of the global economy have come together to manage the "climate" through the "cloud." The recessionary financial and business climate—like the cold, windy, rainy weather that was posited as a benefit for data-center cooling—was perfect for the "logistical media" (Rossiter 2016) of data centers, nourished by proprietary subterranean flows of capital from abroad, which generated these large, power-hungry, and labor-averse infrastructures. Entangled with the cultural and environmental politics of the country through data centers, the state and capital frequently employed cultural rhetorics to naturalize the role of big tech companies in the infrastructural provision not only of capital and connectivity but of future care through industrial employment. Representing a fulcrum of public and private partnership, where the state has bent over backward to ensure the proper infrastructure, planning provisions, technical labor, and business environment to facilitate these massive and power-hungry facilities, data centers represent the coalescing of various logics at the heart of Irish spatial development in the post-crisis and recovery era. Through this array of conditions these companies and their partners in the state engage in what I call "climate extraction": a coalescing logic of extraction that can only be achieved in partnership with the Irish state's persistent association of the country's space and labor with "greenness" and the naturalization of the infrastructural conditions of an FDI-driven economy tethered to global markets. Looking at both the strategies of the state in attracting data center investment and the politics surrounding how people relate to them as potential and existing infrastructures that represent an imagined future "stability," in this chapter I try to understand how cultural, political, and environmental histories and presents are not necessarily existing in easy antagonisms between communities/workers and the state/capital, but rather in a far messier set of relations, requiring a reorientation of our conceptual tool kits for doing politics in the era of big tech's media infrastructural dominance.

Conclusion

As this book takes the reader in and out of a variety of interconnected but differentially globalized worlds across Ireland, I hope that the dynamic and lively environments encountered throughout are generative for understanding how the cultural and mediating relations of capital have reorganized in an era of rampant and naturalized austerity. The machinations of global finance, tech, and media capital spare few territories their omnivorous and extractive operations, and Ireland, as throughout history, has been an experimental case where particular kinds of media infrastructural activity were tested through the conditions of austerity pervading through the financial crisis—a neoliberal "wild west" where these turbulent circulations have formed monstrous new infrastructures amid the disordered foundations of the old. However, this book makes clear that different political futures arising out of the networks of financial globalization and the dense entanglements with life and culture in Ireland are in fact possible. But to disrupt the commonsense logics that are driving the processes of media infrastructure in Ireland and elsewhere, we must first learn how they operate and become intimately acquainted with how they are reproduced.

1 —— Turbulent Waters

Media Infrastructure and the World Economy

Ireland's early experiments with liberalization under Seán Lemass were among the first steps toward becoming a proactive nation-state participant in the world economy. As world systems theorists such as Giovanni Arrighi (2010) and Immanuel Wallerstein (1976) have articulated, the interconnections of the global economy subject diverse locations across the world to the fluctuating cycles of accumulation and financialization promulgated by very powerful productive centers. While the United States has dominated the current cycle—what Arrighi calls "the long twentieth-century" (2010)—inherited from the British Empire, the height of each cycle leads, according to these theorists, to deepening crisis and eventual systemic collapse. Many saw the global financial crisis of 2007–8 as evidence of the imminent collapse of US hegemony.[1] But sometimes lost in these global understandings of power and hegemony are the devastating effects of these cycles (and collapses) on so-called peripheral or semi-peripheral countries and regions across the world. Like other European semi-peripheries Greece, Italy, Portugal, and Spain,[2] in the aftermath of this crisis, Ireland was left shattered, and the sociocultural and political formations of the country were still navigating through the aftermath of "recovery" amid successive planetary crises through the late 2010s.[3]

This chapter outlines the history of the Irish state's active entry into the "slipstreams" of the world economy leading up to the Celtic Tiger and 2007–8 financial crisis. In doing so, it also builds a theoretical grounding through which to understand the role of media and infrastructure during the crisis and recovery. While many have differently unpacked the methods by which capital's omnivorous operations occur on a world scale and entangle with specific places and environments (see, for example, Mezzadra and Neilson 2019; Moore 2017; Tsing 2015), this chapter lays the groundwork for how to understand these processes through the unique history and context of Ireland, where exactly they "hit the ground" (Mezzadra and Neilson 2019), and where they are then operationalized or disrupted through media. Media and the media industries are fulcrums through which the global mechanisms of finance capital are mobilized in Ireland, through the historic strength and pervasiveness of coalitions between the state and capital and the mobilization of "culture" within Irish national development. The introduction detailed how these tightly wound tethers mean that when the financial markets rock, Ireland rolls. As global turbulence increasingly becomes the norm, both in markets and in nature, the sites at which the wildly uneven flows of global capital affect local workers, communities, and environments must be analyzed in their *emplaced* implications to push back against the place-agnosticism of finance capital. Financialization comes to structure particular relations from both above and below because of historical, current, and emerging political and cultural conditions. Ireland's status as simultaneously a wealthy, highly "developed" European nation-state and a country with a complex postcolonial history makes it an instructive site through which to study how "neoliberal" systems have mapped onto historical processes and geographies.

Beginning with the history of (neo)liberalization and financialization in Ireland, this chapter builds on formative theories and critiques of "uneven development" in the context of a "semi-peripheral" country like Ireland. After setting up a brief history of this developmental context, I examine the implications of "media globalization" on the Irish media industries. Finally, as a site through which to expand the discourse on media industries and spatial development in Ireland, the chapter undertakes an analysis of Ireland's infrastructure, the built spaces through which goods, information, people, and capital *materially* circulate. Specifically looking at the place of "media infrastructure" in post–financial crisis Ireland, and the role of logistical media and processes of *naturalizing* of these spatial flows and their extractive mechanisms, the chapter excavates the bedrock on which financial-crisis Ireland ran aground and attends to the roles of built space, policy, and technological infrastructure within the recovery strategies of the Irish state from 2007–8 until 2020.

Liberalization and Spatial Governance in Ireland

Neoliberal financialization in Ireland is a story of postcoloniality, liberalization, and foreign direct investment (FDI), historically specific conditions in Ireland that both reflect and complicate existing histories and theories of uneven development and financialization.[4] In world systems analysis, Ireland has often been considered a "semi-periphery" (O'Hearn 2001), not wholly marginal but dependent upon the "core" productive centers and circulatory apparatuses of power and capital in Europe (and the United States). Literary scholar Luke Gibbons notes that the eventual liberalization attributed to Lemass's policies, far from a smooth, even, and progressive process, was a stilted, fragmented endeavor that further ideologically divided the industrialized (cosmopolitan) cities from the rural (nationalist) countryside (1996, 84). Sociologist Denis O'Hearn argues that emplaced specificities of industrial and spatial development across urban and rural Ireland are as important within uneven development debates as the global processes that affect them (2016, 203). As a "developing" postcolony in the mid-twentieth century, Ireland was never an evenly "industrialized" country by European standards. This patchy and unconventional path of modernization, largely agrarian and protectionist prior to the FDI and financialization that would characterize the late twentieth century (Deckard 2016), contributed to the ongoing separation between rural and urban political and labor organizations. Both within and outside of Ireland, along with financialization came the unloading and exporting of risk onto increasingly precarious classes of people, workers, and environments along familiar paths of exploitation and marginalization.

Ireland's entry as a semi-peripheral participant in the slipstreams of the global economy is a generative example of the ongoing resonances of and problems with "uneven development" theories (Smith 2008), especially in relation to the spatial politics of modernity. The Industrial Development Authority (IDA, today IDA Ireland), the semi-state organization in charge of attracting and maintaining FDI in the state, has long advertised Ireland as an ideal and competitive place to invest, do business, and relocate a variety of operations in their holdings and industrial parks built across the urban and rural regions of the country. Quoting the artist Robert Ballagh, Gibbons points to the yoking of modernity and tradition in the Irish state's developmental ideology during early stages of liberalization: "Those who judge Ireland by its promotional images [from the 1970s–80s] abroad must risk a certain cultural schizophrenia: 'You have the IDA out in the US selling Ireland as a modern progressive go-ahead capitalist society. Invest in Ireland and make a profit. And you have Bord Fáilte eulogizing roads where you won't see a car from one end of the day to the other: it's almost as if they're advertising a country *nobody lives in*'" (1996, 86, my emphasis). For Gibbons, the Irish state prepared an ideological space through which capital could flow unchecked, over unspoiled nature, with no human life to get in the way: "The location of advanced technological factories in remote, often spectacular, settings was motivated, not by a love of the picturesque as the IDA's own copy would have us believe, but by the more prosaic imperatives of a regional policy which guided the IDA's industrial strategy until the early 1980s. This policy could be described as industrialization without urbanization. Part of the attraction of outlying rural areas for industrial investment was that they lacked the strong traditions of trade union militancy which are characteristic of the urban working class" (88). These interlocking development strategies of deregulation, the disempowerment of labor, and the opening of national space to global capital, at whatever social or environmental expense, has long been activated across *internal* striations and characterize neoliberalism as an emergent strategy for national economic growth based on the attraction of FDI.

Ireland's self-promotion as a "wild west" for multinational investment was largely a state-led developmental strategy responding to successive crises from the 1950s to the 1980s and global market pressures. Postcolonial state strategies in the twentieth century have frequently been labeled "developmental," based on their centralized governing strategies to directly intervene in labor regulation, invest in key sectors of the economy, and maintain a role of basic economic planning in order to remedy an industrialization "gap" perceived as uneven development (Mezzadra and Neilson 2019, 115–32). Across Ireland's postcolonial era, the principle of the developmental state holds true, as the state crafted imperfect strategies of independence in sectors like energy and agriculture (Mercier 2021), which were also characterized by a structural inability to *fully* provide for an entire national population under centralized economic plans in a context of historical underdevelopment and ongoing dependency, first on Britain and later on the United States (Bresnihan and Brodie 2025). One result of this was that swaths of rural areas remained economically and infrastructurally cut off for much of the twentieth century (Bresnihan and Hesse 2021).

This compromised developmentalism was superseded, slowly but surely, by a consensus toward export-led development from the late 1950s onward, which intensified in scale and focus through the 1980s (see Bresnihan and Brodie 2025). In this periodization, the neoliberal measures of the Celtic Tiger and its preceding years are more accurately "post-developmental" (O'Hearn 2001), as state developmental policies adjusted to privilege multinational investment as the *primary* driver of economic activity. The permanent "crisis" of the developmental state, especially due to external economic pressures, left countries like Ireland disproportionately open to neoliberal reforms and financialization, as well as a growth in extractive activities (Mezzadra and Neilson 2019, 124–27), which would come to characterize post-developmental economies (Ong 2006). Ireland's rapid shift from limited and agrarian industry to postindustrial and FDI-led sectors like finance, technology, and pharmaceutical manufacturing was adapted to the increasing influence of multinational industry in the world economy (Ó Riain 2000), plugging into the cash flows of manufacturing via exports (whether through manufacturing or extraction) as a central method of regional and national development.[5]

But this has its origins in the compromised developmentalism of the postcolonial state, from which the post-developmental state inherits its national economic remit. As anthropologist Aihwa Ong notices in post-developmental economies in East Asia, "zones of variegated sovereignty," which characterize export-led post-developmental strategies, break up the space of a nation-state while sharing the role of governance with private, mostly transnational, enterprise (2000). The Shannon Free Zone (SFZ) in the west of Ireland, a tax-free special economic zone (SEZ) established in the late 1950s, pioneered this approach to "developmentalism" in a regional context (for a rosy overview, see Callanan 2000). It also represented an early template for what neoliberal "planning" might look like, wherein the semi-state Shannon Development was given exceptional control over economic measures and even governing control over the town of Shannon until 2004. The SFZ, which maintained its exceptional tax benefits until 2003, became an offshore manufacturing hub for US companies, representing an early example of export processing and supply-chain production.

A similar strategy would be tried again with the establishment of the International Finance Services Center (IFSC) in the Dublin Docklands as an SEZ in 1987. The IFSC tested, with EU permission, an exceptional 10 percent tax rate on the premises, transplanting the export-processing model of the SFZ to Dublin for financial services. As sociologist John Urry notes, offshore financial centers in postcolonial states are compelled by neo-imperial influences to facilitate powerful flows of capital to "forge a new economic position within the emerging global economy" (2014, 50), demonstrating the simultaneously active and reactive role that peripheral and semi-peripheral states play within world systemic relations. By 2010, the IFSC was the second-largest center for offshored finance in Europe (MacLaran and Kelly 2014a, 24). These spatial deregulatory measures were supplemented by property-tax incentives for private-built development in the late 1980s, often centered on the area surrounding the IFSC in the Dublin Docklands (22–24). Along with the location of IDA parks in separated suburban and rural regions (from the 1970s and through today) and designation of demarcated Strategic Development Zones (SDZs) for private contracting in cities (since the early 2000s), these variegated strategies masked the overall expansion of tax, financial, and planning incentives for transnational capital across the whole of Ireland. By the late 1990s and early 2000s, all SEZs were essentially harmonized when the state implemented a 12.5 percent nationwide corporate tax rate, only 2.5 percent above the exceptional IFSC rate and the second-lowest in Europe, effectively turning Ireland into a countrywide SEZ for multinational financial operations. "Tax haven Ireland," as political economists Brian O'Boyle and Kieran Allen refer to it (2021), was born.

These logical shifts in economic strategy, which opened the country to FDI and laid the conditions for rapid economic growth, parallel, if not mirror, the trajectories of postcolonial and post-developmental states in the Global South employing similar measures. Ireland's deployment of deregulated zoning and bordering mechanisms, from the SFZ, the IFSC, to SDZs, to nationwide corporation tax strategy, for example, should be compared with those of the East Asian "Tiger" economies studied by Ong (2000, 2006) and to which the Celtic Tiger owes its name. Through the development of SEZs and free trade zones, power relations on a global scale were reproduced locally, multiplied across sequestered zones, and given the appearance of a seamless, "smooth" world despite the patchwork territorial layout of production and circulation. Ong, when describing her theory of the "graduated sovereignty" of such SEZs in East Asian post-developmental economies, notes that post-developmental strategies involving "patterning of production and technological zones . . . designed to facilitate the operations of global capital" meant that "national sovereignty" tended to entail the redirection of capital, technology, and expertise into strategic areas *defined* by the operations of global capital (2006, 78). This made labor and spatial exploitation *infrastructural* within even publicly defined strategies of growth and development. As Ong continues, states thus "adjust political space to the dictates of global capital, giving corporations an indirect power over the political conditions of citizens in zones that are differently articulated to global production and financial circuits" (2006, 78).

What is essential for Ireland about Ong's argument, beyond the comparative utility of analyzing public/private alliances to make states or regions "competitive" for global capital, is that strategies of zoning *for* capital's operations—across manufacturing, services, and knowledge production—manage regional and national labor pools via intentional segmentation and subjugation to regimes of capital accumulation. While Ireland may not administer the degrees of exploitation and inequality characterizing other post-developmental and extractivist governments in the Global South (Gago 2017; Gómez-Barris 2017; Ong 2006),[6] the state continues to imagine its own internal spaces via this intensified focus on inward investment that treats space as a source of value extraction, with things like culture and labor emerging as either obstacles or fuel for more expansive operations and participation in uneven global networks.

Sovereignty and territory transform under these transnational arrangements of trade and finance—or what architectural theorist Keller Easterling refers to as "extrastatecraft," whereby "multiple, overlapping, or nested forms of sovereignty" can be observed "where domestic and transnational jurisdictions collide" (2014, 15). Through these spatial sovereignties of transnational capital, basic notions of democratic consent and dissent are challenged via spatial technologies and bordering mechanisms that carry a primary form of capitalist coercion. As the above-mentioned 12.5 percent corporate tax rate was harmonized across the country and special exemptions have been given to intellectual property (IP) and research and development (R&D) profits (which often grants effective tax rates of less than 5 percent), the national exchequer has become a site of frequent conflict over national development strategy. Multinationals exert substantial power over Irish economic decision-making due to the scale of investment, and the unpaid tax bills of massive tech corporations such as Apple (see McCabe 2022) have led to friction between Ireland and supranational regulators. According to the EU, in the late 2010s, Apple owed the Irish state €13.1 billion due to what are viewed across Europe as unfair tax advantages. Ireland refused to collect this tax bill for fear of losing these tech giants, even appealing the EU's ruling on the matter, not touching the assets transferred by Apple.[7]

Thus, while attracted to Irish shores by exceptionally low corporate tax rates, especially on IP and R&D (central to tech services' knowledge economy), big tech corporations like Apple have been lured into staying by a state and planning environment that enthusiastically facilitates their extractive transnational operations and infrastructures. In Ireland and elsewhere, rather than flattening out regional and national borders, globalization initiated a spatial order that operates across differentiated and "competing capitalisms," with each having their own "specific local, regional, and historical contingencies" (Lee and LiPuma 2002, 205). The erosion of hard-won workers' rights across these eras is largely the result of such races to the bottom in a disorganized, disaggregated, and arguably "post-Fordist" global landscape (Hardt and Negri 2000; Lash and Urry 1987), by which these geographical contingencies also mold links to wider productive relations.

However, while post-Fordist analyses frequently lament the loss of worker-centered democracy, with a global lens, we should emphasize: Industrial capitalism as it is known in the Global North is an exception on a world scale, as Brett Neilson and Ned Rossiter argue (2008). While mostly white workers in core countries enjoyed certain kinds of stability across the twentieth century, labor precarity and the exporting of imperial violence to the peripheries has remained an ongoing and defining characteristic of production and circulation in the capitalist world system—and has shaped the relations and strategies of development outside of the "core." This reality has frequently been obscured by the serendipitous centrality of Fordism in industrial production during high modernity in the early twentieth-century imperial core. Anthropologist Anna Tsing argues a similar point in relation to emergent precarities across a variety of political economies and ecologies, asking "what if precarity, indeterminacy, and what we imagine as trivial are the center of the systematicity we seek?" (2015, 20).

Ireland, as a semi-peripheral, postcolonial country that has wagered its prosperity on one of the most liberal financial and industrial regimes in the "industrialized" world, offers an edge case through which to analyze the specific industrial and developmental forms that have arisen in a "post-Fordist" industrial landscape of extreme austerity, in a place where industrial capitalism has never looked as uniform or as even as theories pioneered in the imperial metropoles would suggest. Since the 1980s, however, the outweighing influence of multinational capital, and a regulatory regime happy to let tech and media companies work through and manage the ongoing contradictions of growth, mean that Ireland has once again become a kind of laboratory for a paradigm of managing, innovating, and profiting from turbulence. Centering change, imbalance, turbulence, and precarity within global systems, rather than fixity, balance, stability, and security, complicates these teleological narratives of progress and modernity across historical landscapes writ large—especially in the context of market fixes and solutions to constantly unfolding crises.

Circulation, Turbulence, and "Neoliberalism with Irish Characteristics"

Neoliberalism, along with the historical processes that preceded its implementation in Ireland, tethered Ireland to the fortunes of global finance, moored in the turbulent waters of the world market. As a market ideology whose spatial and value calculations have defined Irish economic development since at least the early 1990s, neoliberalism is not only a category of governance or a theory of development. Rather, it is a materially specific institutional instrument methodically integrated into economic common sense across the world. The subsequent development of global infrastructures of financialized trade has conditioned particular subjective responses to the global market and come to define political responses to its circulations. This includes deregulation and strategic zoning of space at the state level as much as the popular and social disinvestments from state care and the commonsensical market rationality that dictates everyday relationships to things like infrastructural provision and welfare (see Haiven 2014).

Marxist geographer David Harvey situates the beginnings of the neoliberal era from the onset of the 1970s, following the slow creep of neoliberal economic thinking into the global political mainstream. This was engineered by a semi-organized contingent of economists (including Friedrich Hayek and Milton Friedman), known informally as the "Chicago Boys," who espoused the principles of neoliberal economic theory, especially its deregulatory measures, through university programs and well-funded think tanks throughout the mid-twentieth century, reaching practical policy implementations in the United States and United Kingdom by the mid-1970s (Harvey 2005, 20–23). The subsequent "disorganization" of global capitalism (Urry 2014, 29), echoing the fragmentation and supposedly liberating messiness of an emergent "postmodern" epoch, coincided with the rise of supranational organizations such as the World Bank and the International

Monetary Fund (IMF), the growth of floating exchange rates and money markets, the expansion of global telecommunications technologies, and algorithmic trading.

These processes systematically incorporated "external" markets into a global funnel of financial accumulation, via the correlating structures of existing imperial power networks and the orthodoxies of neoliberal economic strategies, which continues to siphon goods and profits into the rich Global North while amplifying inequality and subjugation in developing regions of the Global South (Urry 2014, 175). As sociologist Melinda Cooper describes the situation, "The integration of financial markets in the early 1970s initiated a period of enduring and structural turbulence in world economic affairs . . . the unpredictable was, of necessity, factored into the calculus of world economic futures. Henceforth turbulence could not be prevented; it could only be managed" (2010, 167). Any modicum of economic planning in neoliberal nation-states transitioned into defanged management of global turbulence in the market (which is often accompanied by political instability, unrest, and outright revolt). The swings of this turbulence are felt more strongly in the so-called developing world (Cooper 2010, 168), as sociologist Campbell Jones argues that "capitalist power has of course rarely been satisfied to simply let the market do its magic, but rather has always carefully planned the extension of market forces, which it then withdraws or modifies when they are not achieving the desired ends" (Jones 2020, 3). However, the decline of liquidity in US funds meant that, according to Cooper, "turbulence that could once be safely exported to the peripheries of the world economy has now returned to haunt the heartland" (Cooper 2010, 169). These periodic shifts and the migration of capital to new markets have restructured "uneven development" debates and new forms of national and regional competition under financialized capitalism (Harvey 2005, 87), structured as these economies are by the turbulent circulations of finance.

Ideas of uneven development, center/periphery, Global North/South, have traditionally relied on geographical, ideological, and temporal divisions from an ideologically northern perspective. Undoubtedly, neoliberal measures structuring the global economy are overwhelmingly centralized in the Global North, and (neo)imperial power is exerted from productive centers and the "global cities" of finance (Sassen 2001). But these hierarchies, while tipped toward these northern financial centers, are reproduced locally across the developing and developed worlds as the value calculations of neoliberal capitalism, running out of "outsides" for productive expansion, turn inward and intensify in all places, in often imperceptible and apparently mundane ways. Political economists Benjamin Lee and Edward LiPuma write throughout their work about the cultural imaginations of financial circulation, discussing how the naturalization and "collective agency" of the "market" proceeded as "circulation-based capitalism" became the dominant global economic model (2002). Karl Marx argues that "circulation" and "markets" can only occur once the "free worker" finds themselves within a "labor-market . . . as a particular branch of the commodity market" (1992, 273). This is formative for Marx, especially because this is in no way a "natural history," but rather the product of a historically specific set of circumstances, of "many economic revolutions, of the extinction of a whole series of older formations of social production" (273). However, the collective imaginations of labor's subjugated and dependent place within a "global market" somehow remain constant even as so-called social totality, and circulation with it, change with progressive cycles of capitalist accumulation and modes of production. Lee and LiPuma update Marx to account for the circulation-based capitalism of financial globalization in the way that circulation's abstraction of labor constitutes an "internal dynamic that, in its 'unfolding' or 'self-positing,' self-reflexively constitutes itself" (2002, 199). This turn inward in the absence of a necessary outside requires a series of naturalizations of dialectical binaries and power relations through which all aspects of social life can be subsumed under the totality of exchange value (Lee and LiPuma 2002, 203; also Smith 2008), driving the total imaginary of the "market" and the constitutive place of labor as a source of value within it. In effect, circulation itself becomes "the cutting edge of capitalism" (Lee and LiPuma 2004, 9) in contemporary global networks.

But the "turning inward" recognized by these theorists points to crucial elements of financialization within political subjectivity, and is central to what I refer to throughout this book as the "naturalization" of global capital's value dynamics in and through Ireland. As neoliberal capitalism has intensified embodied value relationships, so have workers' and communities' lived experiences of systemic fragility and turbulence, throwing into relief the structural precarity that is exported to the margins of ongoing uneven development. Financialization and the attraction of FDI across the whole of Ireland, along with a series of deregulatory measures in the late 1980s, conditioned early neoliberal Ireland for more efficient production within global systems of capital, while under the surface dismantling social and economic protections. Between the IFSC's establishment in 1987 and the 2007–8 financial crisis in Ireland—the limit points of the Celtic

Tiger—faith in "light-touch" neoliberalism grew as foreign investment ballooned in the financial, tech, and pharmaceutical sectors and GDP per capita surpassed the UK's (MacLaran and Kelly 2014a, 25). Rose-tinted memories of prosperity and chimeric affordances of middle-class luxury unevenly distributed across the country were buttressed by the persistent consensus of the "social partnership model" across the state, industrial interest groups, and unions, which had "effectively denied the essential conflict of interest between labour and capital and obscured the capitalist state's fundamental role of securing conditions favourable to capital accumulation" (MacLaran and Kelly 2014a, 26) through the 1980s and 1990s. This consensus among Ireland's middle classes toward FDI-led prosperity helped mask the predatory flows of global finance operating underneath the model, its fragile structures held together by cheap materials and superficial wealth.

Thus, by facilitating the circulation of offshored finance and its logics of value, Ireland's social and political fabric was made more receptive to the "rule" of capital and its ability to provide jobs and welfare, demonstrated by the widespread prosperity and excitement of the Celtic Tiger as much as by the exhausted acceptance of neoliberal reform and austerity measures after the financial crisis. Not seeking to conflate the differing but interlocking histories of neoliberalism and financialization, however, I want to emphasize that the heightened forms of financial accumulation brought on by a neoliberal world order across and through nation-states instituted the creep of the finance industry's value calculations into economic common sense across the world. Adapting Moore's dictum that "Wall Street is a way of organizing nature" (2011, 43), literary scholar Sharae Deckard similarly contends that "the IFSC is a way of organizing nature, with pernicious consequences for water, energy, and food systems in Ireland" (2016, 158). Its role in establishing and continuing Ireland's function as a tax haven has expansive effects not only across built environments and infrastructures, but on the very ground and resources through which people subsist. Neoliberalism with Irish characteristics means the integration of the state and civil society into a consensus of a national character determined by multinational investment and modes of financial facilitation.[8]

Financialization, as cultural theorist Max Haiven argues, refers not only to the finance industry itself, but also to the pervasive sociocultural implications of financial dominance across the world and in everyday life (2014). For example, in her work on Wall Street bankers and the finance industry, anthropologist Karen Ho identifies the implications of *living* financial rationalities (2009). Not discrediting how powerful the financial industry was prior to the financial crisis—and continues to be today—she penetrates the everyday and lived ideologies that reproduce market goods and financialized subjectivity as a cultural formation and activity somehow parallel but influential to actual financial trading and services. These "institutional cultures and arrangements as well as expert evaluators of spokespeople for the market . . . can give us insight into the workings of 'the market'" (Ho 2009, 184–85). With the power of finance so vast, the everydayness of finance cultures seems a small-scale factor within its global reach. But the institutional and risk-taking tendencies of finance—as well as the gendered, racialized, and meritocratic structures of the industry itself—have enormous implications for the world economy, contribute to its structural turbulence, and are felt (and reproduced) on an everyday basis across a variety of scales.

This ideological reproduction in industrial contexts defines how certain norms come to affect environments within which decisions and transactions occur within a market, and this happens to different degrees at all scales: whether in boardrooms, coworking cafés, or on the job market. As political economist Christian Marazzi reminds us, "The financial economy today is pervasive, that is, it spreads across the entire economic cycle . . . finance is present even when you go shopping at the supermarket and use your credit card" (2011, 27). We thus need a hybrid approach that understands the pervasive ways that its logics operate in necessarily multiscalar ways through spaces and communities. Ho's focus on ground- (or high rise-) level phenomena allows insight into how the larger systems work, continue, or break down. As Ong notes, culture is not "somehow separated from 'rational' institutions such as the economy, the legal system, and the state" (2000, 23). The "culture of financial circulation," as Lee and LiPuma argue, describes the very integration of life and labor with financial rationalities. Across the timeline of (neo)liberalization traced in this chapter, different cultural logics have developed in relation to these economic realities that have shaped the contemporary spaces and environments of Irish media infrastructure.

Just as any understanding of financialization needs to be attuned to the transcendence of scales, so too must it keep space at the center of discussion. Finance's turbulence is most acutely experienced through the material environments that it creates, manages, and destroys. Lee and LiPuma puzzlingly argue that financial instruments are largely detached from the sites of their greatest effects and their "abstract symbolic violence" (2004, 28). However, on the contrary, space, and built space especially, is immanent to how financial mechanisms operate, its logics spread, and its violence is enacted. Ireland's spectacular property crash and the widespread ruin of its aftermath demonstrate the degree to which any statement about finance's "symbolic" detachment must be filtered through a concrete understanding of its spatial mechanisms and the internal logics through which financial machinery operates through property. For example, geographers Andrew MacLaran and Sinéad Kelly describe the use of Integrated Area Plans (IAPs) (2014a, 31) as well as SDZs in Dublin designed to "streamline" development (Murphy, Fox-Rogers, and Grist 2014, 55) in the city and its suburbs in the years following the Celtic Tiger. Property tax incentives and the deregulatory environment left Ireland doubly vulnerable to market fluctuations. In the 1990s and 2000s, these and the other spatial and deregulatory measures listed above effectively "tied the fortunes of [Dublin] to the highly cyclical performance of property-development markets" (MacLaran and Kelly 2014a, 34), directly and inextricably entangling spatial futures (and ruin) with financial speculation.

When the markets crashed in 2007–8, of course, these financial and supply-chain entanglements of the world economy set off a wave of contagion that spared few territories internationally, especially in Europe (Marazzi 2011). Capital either evacuated or dried up, leaving visible economic devastation and spatial ruin in its wake. As the knock-on effects of debt defaults and toxic assets spread from financial offices in New York, London, and Frankfurt to the coffers of banks in Ireland, the panic led the Irish government in September 2008 to guarantee the six major Irish banking institutions before recognizing the full extent of the crisis (Coulter 2015, 7). Economic recession at a world scale, and across other "semi-peripheral" European nations such as Greece, Italy, Portugal, and Spain, led to mass unemployment, emigration, and economic stagnation, which would continue through the crisis. By November 2010, after sinking astronomical sums of public money into the banks, including €30 billion into Anglo alone, the government revealed that it was negotiating terms of a bailout with a coalition of international institutions, which would come to be known as the "troika" across Europe: the International Monetary Fund (IMF), the European Commission, and the European Central Bank. The terms of the bailout were incredibly lucrative for these institutions, although the conditions for Ireland, as Colin Coulter points out, would have been referred to as compulsory "structural adjustment," typically reserved for postcolonial nations in the Global South, had Ireland not been one of the world's wealthiest countries per capita by this point (2015, 9).

The program of austerity that followed, which forcibly slashed public jobs and wages, eroded previously untouchable social services, and cut public funding across the board (Coulter 2015, 9), was catastrophic, leading to a period referred to as a kind of national "soul-searching." For a brief period, it appeared as though alternative futures beyond neoliberalism and FDI might be possible for the small island nation (see Linehan and Crowley 2013; O'Callaghan, Boyle, and Kitchin 2014). However, unlike in Greece and Spain, this ostensible structural adjustment was not popularly protested beyond a few specific issues such as water charges (see Bresnihan 2016), demonstrating the effectiveness of the Fianna Fáil government's austerity rhetorics of "it's our fault" (Hearne 2013), the successful disempowerment of labor union leadership (Coulter 2015, 10), and the related pervasiveness of neoliberal common sense within how the Irish middle classes saw the country's economic fortunes as individualized and not necessarily inter-related to a world system that systematically and cyclically punishes the indebted (Lazzarato 2012). This was true even as unemployment reached a peak of 15.5 percent in 2013, which paradoxically would open space for arguments about multinational jobs leading regional recovery programs (see chapter 4). What Ong calls the trademark "disciplinary neoliberalism" of IMF structural adjustment protocols (Ong 2006, 93), imposed by debt, transferred economic sovereignty to supranational regulators and further removed the country's economic fate from the hands of ordinary residents. The wave of coerced privatization that followed continues to this day. It is often enacted through the contracting, leasing, or selling of public services and assets to for-profit institutions and multinational corporations, as well as through more back-door marketization measures such as financial incentives for infrastructural and industrial development and operation such as smart meters and other individualized measures framed as "cost-saving."

Thus, the cycles and interdependencies of the world market and its eventful waves of circulation and accumulation left Ireland in the wake of financial crisis depleted and dejected, subject to a period of brutal austerity during which labor was further disempowered, space deregulated, resources and services privatized, and structurally transformative solutions left off the table. Further, as I am suggesting, it also laid the groundwork for more speculative and experimental projects of multinational innovation. Rather than adjusting course and seriously interrogating the value of this system of FDI-driven development, the Irish state doubled down and forged ahead with a strengthened neoliberal consensus (O'Callaghan et al. 2015). To do so, it moved into the "recovery" by simultaneously deregulating space and redeveloping Dublin (especially via the National Asset Management Agency [NAMA] and its vulture fund partners), persistently slashing or conditioning cultural funding with profit-making metrics and incentives, actively coaxing the tech industries, and in doing so increasing their power over spatial development. As suggested at the start of the chapter, things like tax incentives have transformative effects on space as *infrastructural* to capital circulation, and have led to mounting contradictions in the public life and media of development in Ireland. Flows of finance travel across and affect territory via such state policies, which create the conditions necessary for the privileged movement of capital through the built environment.

Emplacing Media Globalization

The long decade of austerity in the 2010s has reshaped the media industries in Ireland (and across the world), rife with precarious work, short-term contracts, and the overall disempowerment of labor at the hands of an increasingly privatized, conglomerated, and FDI-driven set of media and cultural policies. While labor has long been a primary concern of media industries research (see Caldwell 2008; Mayer, Banks, and Caldwell 2009), the financialization of the media industries has come to be a greater topic of analysis (Caldwell 2023; DeWaard 2020; Kidman 2019). But the specificity of its interaction with "media globalization" in places like Ireland demonstrates how financialized media industries materially operate: breaking down infrastructural barriers to capital accumulation by establishing new pathways and dismantling empowered labor (Curtin 2016). The restructuring of the global economy via neoliberal globalization has seen media and its technologies become a fulcrum through which information and capital travel, for example, through communications technologies, the basis for Harvey's famous characterization of "space-time compression" under financial globalization and its communicative apparatuses (1990). These shifts have come to implicate both media production and distribution, and thus national and localized industries as much as the global distribution of capital, labor, talent, and content through the cooperating state and private, spatial and financial mechanisms traced in the preceding sections.

These are interwoven and interdisciplinary fields of examination. But through textured analyses of media industries and media policies in the small, extremely "open" economy of Ireland, especially while also examining the wider supply chains and production pipelines of the global economy, the material implications of media globalization—and its particular national and transnational arrangements—come into much sharper focus. As financialization extended into the very fabric of how we live, work, and relate to one another, financial and logistical rationalities and practices came to shape localized industries and spaces beyond traditional media encounters.

Building on the previous section, capital's diverse entanglements require centering "circulation" as the *core* tenet of contemporary capitalism, and this includes within media production apparatuses. As this book argues, it is the politics of *circulation* that define not only how media moves *after* it is produced in the form of distribution through cinemas, television, and digital platforms—in short, how it enters the eyes, ears, and homes of consumers, as has been extensively analyzed by scholars of distribution (Curtin, Holt, and Sanson 2014; Lobato 2012; Perren 2013). However, circulation, in a Marxist framework, also influences how media industries and their infrastructures are reshaped by global circuits of finance, logistics, and supply chains. The study of infrastructure is the study of the physical spaces and networks of circulation (see Larkin 2013)—and in media industry studies, this means shifting focus from media production and distribution *specifically* and into an analysis of the "noisy sphere" of circulation and its constitution of labor markets, modes of production, and supply chains (Marx 1992, 270–80; also Lee and LiPuma 2002, 199). After all, production and productive sites, whether in film production or device manufacturing, are beholden to the dictates of global supply chains (see Mayer 2011). This includes, for example, cheap media production by precarious migrant workers in free trade zones in Abu Dhabi (Dickinson 2024), the location and gutting of VFX companies servicing Hollywood films in Canada (Acland 2018), or the local laborers recruited to assist with *Star Wars* production in rural west Kerry (see chapter 3). These diverse arrangements of labor—from mushroom pickers in the US Pacific Northwest (Tsing 2015) to television factory workers in Manaus, Brazil (Mayer 2011)—illustrate the flexible, diverse, and decentralized nature of global economies, and the precarity and violence that this introduces (and intensifies) for workers and livelihoods across the media supply chain.

Film and media studies has sometimes avoided the direct conflation of film and capitalism as an inextricable relationship, perhaps due to ongoing debates around research emphases on film and media as "art" or "industry" (see Holt and Perren 2009). However, between the industry's imbrication within multinational digital systems of content and capture and the long history of exploitative capitalist media-production practices worldwide, the necessary reappraisal of the medium's role in and contribution to global networks of capital is long due (see also Grieveson 2017). As the labor pools of creative workers—"disposable human material" (Marx 1992) for global production and services—grew in Ireland and elsewhere through the 2010s due to increased training, industrial opportunities, and labor migration, capital (especially multinational media conglomerates) itself migrated to the most profitable centers of production (Curtin and Sanson 2016, 7), whether in search of tax breaks, cheap labor, infrastructure, or location needs (or some combination of these) (see Dickinson 2024). If local and supranational governments provide the proper incentives and exploitable fixed capital and resources (facilities/tech and workers), this process will continue unless labor solidarity and organization respond to the current transnational arrangement of diverse manifestations of finance capital (Curtin and Sanson 2016, 7). As media scholar Michael Curtin articulates (2016), these policies around media built from "creative industries"–focused policy that characterize the "creative city" and its spatial development, as will be discussed in chapter 2. In short, the creative city, by spatializing the private industry–driven cultural logics of creative industries, integrated the creep of privatization into policies of arts and culture, government, and semi-state organizations through the spatial logics of private finance (see Hesmondhalgh and Pratt 2005).

As Irish media scholar Aphra Kerr tells us, in studying media industries within these frameworks, our focus should remain "relational and, in terms of cultural production, it means that we need to investigate how and in what ways industries, companies, workers, texts, and users get embedded in particular spaces while simultaneously being involved in global flows" (2014). Ong describes the funneling of expertise, labor, and technology into strategic sites where the state *facilitates* the mandates of global capital (2006). Digital capitalism and media globalization became intertwined not only through material infrastructures but also through knowledge and social networks by which their conditions of possibility were maintained (see Preston, Kerr, and Cawley 2009), and this included through the protest or acceptance of working conditions. Chapter 3 argues through particular local case studies that the global "race to the bottom," as these competitive regional measures to attract production and creative work have been called, has generated a diverse and competitive landscape of local incentives to attract foreign media production, and labor pools and populations relate to these global systems in a variety of ways, just as the institutions and regulators of the systems themselves are tasked with managing "human capital." What is crucial to this study is that media policies, negotiations, tax structures, and other mechanisms and infrastructures exist not only in the direct production of commodities for consumption. Rather, they organize the "noisy sphere" of transnational circulation by diverting financial and trade flows, encountering and managing local "frictions" along the way (Tsing 2005). While production and circulation are not necessarily mutually exclusive categories even in traditional Marxist conceptions, what needs to be stressed are their deep imbrications and coevality. The constituent role of circulation within production, and circulation's direct production of value, is key to the extraction of value across distances that orchestrates the global structure of film and media supply chains.

The arrangements of global media supply chains are key to how this book understands Ireland's place within the global production and circulation of media content and capital. The Irish Film Board (IFB), historically Ireland's national film body, has in recent years rebranded as Screen Ireland (SI) to reflect its widening remit as an industrial facilitator across a wider range of "screen" content industries like new media and gaming, affiliated with tech and digital media as much as traditional film and television. The country's generous tax break for foreign production (the Section 481, a 32 percent relief scheme which in the late 2010s introduced a yearly graduated 5 percent rate for rural productions) has attracted multitudes of global productions and coproductions on a variety of scales (big Hollywood studios to European "independents") and across the country's diverse landscape, production, and postproduction infrastructures. These tax schemes, however, mean that Ireland becomes one point within wider media supply chains of finance, talent, materials, and information. This has implications for the different places in which these productions actually happen and where "locations" are chosen—whether in rural west Kerry, a studio in Wicklow, or an editing bay in Dublin.

The creative industries in Ireland have been central to the reorganization of culture and media industries under globalization, especially within urban economic development and spatial planning post–financial crisis (Lawton, Murphy, and Redmond 2014). But while Julien Mercille (2014) and others have analyzed the precise role and complicity of public and private media institutions in failing to adopt a critical stance to the Celtic Tiger housing bubble and the endemic warning signs of property market collapse, the ways in which the broader Irish media industries contribute to and participate within certain spatial and territorial practices have only intermittently been examined (see Ramsey, Baker, and Porter 2019; Van Egeraat, O'Riain, and Kerr 2013). Although creative industries logics are holistically applied in strategic planning and development in both policy and spatial contexts (Lawton, Murphy, and Redmond 2014), media (and its labor) are *specifically* mobilized, or slotted into, large-scale spatial and economic plans through their ability to contribute to capital circulation through the country. National media industries, especially after globalization, are not simply factories for producing film and television representations of place and nation. Rather, they also constitute the broader machinery of representation, from film and TV to design and marketing. Media industries, when we look at the more expansive definitions offered by industrial policy and lobby organizations like PricewaterhouseCoopers (2016), include the production of promotional materials, the marketing practices of media organizations, and other outward-facing and future-driven development imagery (see O'Mahoney and Lawton 2019; Rose, Degen, and Melhuish 2014), a whole range of productive activities and their visual culture that promote and normalize future visions of space for capital.

Media, and film in particular, has historically played a huge role in Ireland's presentation of itself to the world (Pettitt 2000)—both intensively and extensively. However, this has largely been focused on how either foreign cinema or "Brand Ireland" has promoted Ireland toward particular cultural or economic ends. Most film productions before the first IFB (established in 1980) were inward productions, often facilitated by the intermittently nationalized Ardmore Studios in Wicklow, which was specifically established in 1958 for Hollywood-style productions during Lemass's tenure as minister for industry and commerce (see Barton 2004). British genre films like *The Spy Who Came in from the Cold* (Martin Ritt, 1965) and *Zardoz* (John Boorman, 1974) were shot at the studios and in surrounding locales in County Wicklow, and much of the nation's film infrastructure and labor resources are still centered in Wicklow and nearby Dublin. However, pre-IFB production in Ireland was most prominent and recognizable globally when it featured rural, idyllic locales, with studio films like *The Quiet Man* (John Ford, 1952) and *Ryan's Daughter* (David Lean, 1970), shot and set in the west of Ireland, receiving international attention. Such films, part of much longer cultural visions of the "wild Irish" coming from England and the United States, paint Ireland as a place of backward and regressive people and politics, whether positively or negatively romanticized as a place of quaint simplicity, violence, or romance. The global "imagination" of Irish culture and landscapes expressed through such films, the "Emerald Isle" trope, has been a straw person within dominant studies of Ireland as a "national cinema," an action that frames the indigenous industry in opposition to the peddlers of trite and stereotyped representations (Barton 2004; Crosson 2003; Mc-Loone 2000, 2008; Pettitt 2000; Rockett, Gibbons, and Hill 1987).

But long-term efforts to unpack the regressive representational politics of these films have tended to turn focus away from complex processes of commodification and broader economic structures at play within popular media. After all, the above-mentioned films were offshored productions, and have nevertheless been used domestically to promote tourism and development in Ireland since they were made. Visual culture, media stereotypes included, has been central to economic development and how Ireland has promoted itself abroad, as a green, wild, but welcoming and business-friendly society, within which the stereotypes maligned by film and media scholars were globalized through processes both intensive and extensive.

In this way, it is instructive to align these representations with those of Irish soft power institutions, in particular the institutional facilitators of foreign capital investment and tourism. Like Gibbons's suggestion that the IDA used Ireland's rurality to draw business (1996) and historian Kevin J. James's study of the commodification of land via rural tourism (2014), literary scholar Anne Mulhall critiques what she calls "Brand Ireland" since the Celtic Tiger, a promotional state that commodifies the nexus between Irish neoliberal (post)modernity and a so-called "premodern" or precolonial past: "borders between heritage and spiritual tourism are particularly indeterminate in relation to the marketing of Ireland, at least in part because the 'Celticity' that is a staple of the Irish tourist industry constitutes Ireland as a therapeutic landscape so that the appeal to the 'mystical and the spiritual' is fundamental to Ireland as a global brand" (2013, 147). She notes that "even as the economic boom was in the ascendance," Ireland was represented as "a refuge from modernity" (Kneafsey, quoted in Mulhall 2013, 147) in Bord Fáilte (now Fáilte Ireland, the state tourism body) advertising. So, while global economic flows have driven spatial development and structured Ireland's domestic industries, most attention at this point in the spatialized study of Irish film and media industries has been paid to aesthetic and cultural developments in relation to "national" artistry, landscapes, and industry. Relationships of these cultural forms to the spatial-financial flows of global capitalism need industrial positioning.

There are thus different stakes within studying Irish media and visual culture at a variety of scales. Corporate planning materials and the visual media of spatial development are tools by which capital brings emergent spaces and avenues of accumulation into being, while also offering a unique vantage point through which to understand the cultures and ideologies of capitalism as enacted through the built environment. As media scholar Joshua Neves argues, spatial planning in urban environments has its own visual culture, which promotes future visions of space and life in the city through idealized digital renderings and public-facing planning materials (2013). These sorts of images and representations are transformative in and of the past, current, and future spaces that they signify. They are financial and logistical visions of future space, which prepare territories for upheaval and extraction. And they are also produced by media design professionals. Promotional and state media using these tools is thus, in this way, interlocking with the media policy agendas they promote, creating imaginative environments through which capital will circulate and extract value from Irish space. Media is not only used as a tool for promoting certain projects and economic goals; rather, it plays an active role at several levels, and encompasses many facets and scales of social organization, reproduction, and labor.

The collapse of "culture" and "creativity" within Irish cultural policy and debates has amplified the existing spatial unevenness of Irish media production, and demonstrates the degrees to which economic common sense has entered the entire structure of media and cultural governance. Diverting this focus to industry and policy practices, like the branding and management of Irish labor forces through "talent" and "skills" and the normalization of precarity in the industry, can carefully integrate these discussions of cultural production into the material conditions driving spatial development. In this way, it is also crucial to expand ideas of the "media industries" to encompass these apparently peripheral media industrial formations that are central to both spatial development and the general employment of trained media and other "creative" professionals. As film scholar Kay Dickinson argues convincingly, the pipeline from university education to precarious media workforce is well oiled (2024), but perhaps more attention needs to be paid to the compromises that media workers make to make a reasonable living. Videography, corporate promotional media, event filming and photography, advertising, educational film—all these areas employ an enormous proportion of the emerging media workforce (see O'Brien, Arnold, and Kerrigan 2021; O'Hagan, Murphy, and Barton 2020). The point is not only that this media drives development, but that the subjects of this media exist across the supply chain, all the way to the point of delivery. The constant and eventful changes in financialized media space have had profound and disruptive effects on the lives and labor of those living in and passing through Ireland, especially through the lens of infrastructure and the labor relationships managed by its spatial inter-sections between policy and industrial practice.

Technological Media Infrastructure Since the Financial Crisis

The built environments through which media industries function are part of wider systems of power and capital. These systems collide with unruly environments, and they manage to operate despite, or "break down" as a result of, friction, disruption, or blockage. In an era of circulatory capital-ism, the principal points of political contention and negotiation are the vast infrastructural networks through which capital operates. As media anthropologist Brian Larkin defines infrastructures, they "are built net-works that facilitate the flow of goods, people, or ideas and allow for their exchange over space. As physical forms they shape the nature of a network, the speed and direction of its movement, its temporalities, and its vulner-ability to breakdown. They comprise the architecture for circulation, liter-ally providing the undergirding of modern societies, and they generate the ambient environment of everyday life" (2013, 328). Study of these built net-works of circulation—and their breakdown in times of crisis, spectacular or ongoing—challenges the apparent "ambient" character of their operations. Workers and communities in Ireland live through the operation and devel-opment of infrastructure differently depending on locations, histories, and politics. Film studios and the "creative city" contain real and affective circu-lations of media and finance; roads and regional development schemes send money and labor across the country to accommodate production; internet infrastructures through which we communicate take the form of private data centers. In each case, space is a tool for value production, superseding the needs of the populace living in it. But through this, infrastructure is also a site of consistent contestation *about* the role of space, culture, and labor within the national narrative of progress and recovery, especially in terms of who benefits from it, who does not, and who suffers for or because of it. Growth and prosperity are unevenly distributed and experienced across the country, and macro-scale measurements such as GDP do not account for these internal economic and geographical striations—especially when it comes to infrastructural provision.

As media scholars Jennifer Holt and Patrick Vonderau articulate, the governance of contemporary corporate media infrastructures tends to oc-clude and "dematerialize" the principles of their operation, disavowing their material consequences while pushing forward strategic forms of visibility

(2015, 81). In Lisa Parks's programmatic "'Stuff You Can Kick': Towards a Theory of Media Infrastructures," she discusses the "infrastructural turn" in media studies, which has seen greater focus on the materials of media: the cables, wires, antennas, satellites, metal casings, anything that is involved in the physical operation of media circulation (2015a). She explores "the sites, objects, and discourses that shape and inform what might be called *infrastructural imaginaries*—ways of thinking about what infrastructures are, where they are located, who controls them, and what they do" (2015a, 355). Through these approaches, she argues, one can read media with an "infrastructural disposition," meaning that media infrastructures, technologies, and their distributive power are made visible in and through the media that they circulate (357). In this kind of politics, silicon and copper intertwine with state and corporate measures to enable both technological advancement and corporate centralization and administer the material and immaterial elements of these "infrastructural imaginaries" whether real or foreclosed (Holt and Vonderau 2015; Larkin 2008; Parks 2015a).

Scholars have long understood that infrastructure constitutes and is constituted by the "publics" that they serve (or don't), especially in terms of the above-mentioned historical unevenness of modernity (Sundaram 2009). Spatial legacies of exploitation, racialization, class division, and other interlocking systems of oppression within capitalism are embedded in infrastructural histories, construction, and operation (Graham and Marvin 2001). Starosielski warns us that the very "friction-free" imagination of media infrastructure has roots in colonial ways of seeing the world and capitalist modes of accumulating capital (2015, 5), which "postmodern" theories of dematerialization have often, however accidentally, fed into. The planning logics of infrastructural development, even in postcolonial contexts, carry active and underlying residues of colonial power and exploitation, even when bringing certain promises of development and modernity to underserved people and regions (see Larkin 2008).

In a resonant example, British railway networks across Ireland, a symbol of imperial modernity, fell to ruin in the postcolonial era, and the country's passenger railway infrastructure remains underdeveloped outside of the Dublin commuter belt (and even within). Transport infrastructure continues to be a point of contention for underserved regions, especially in the rural west, where motorways are sites of both (usually environmental) contention and real or foreclosed dreams of connection and access. The distribution of infrastructural networks—and their direction of flow toward certain strategic points on the map, while cutting off others—has profound effects on the distribution of industries as well as the socioeconomic conditions endured by communities in different places. This is true whether looking at the location of industrial activities, such as how uneven internet infrastructure will frequently exclude rural locations from high tech and postproduction media enterprises, or the basic conditions of everyday life. The fascinating longitudinal documentary *When All Is Ruin Once Again* (Keith Walsh and Jill Beardsworth, 2018), for example, uses the stalled timeline of the M17/M18 motorway in the west of Ireland after the financial crisis to look at the town of Gort at the border of Counties Clare and Galway. Activating a longer cultural and environmental history of "ruin" and cultural relationships to the landscape through a W. B. Yeats poem quoted in the title, the film unpacks the textured public and daily experiences of locals amid their social and infrastructural entanglements with the region's environments, raising questions about who infrastructures are built by and for, what sorts of goods they provide, and to whom.

Infrastructural provision is still anachronistically seen as primarily the purview of states and public services (see Easterling 2014). However, we need to reconsider the equations of public and private, state and corporation, at this advanced juncture of neoliberal capitalism, especially in a context such as Ireland with a long history of public–private partnership (PPP) and semi-state institutions. In Ireland, semi-state corporations and PPPs have been the favored method of infrastructural and industrial development for decades, as private administration of public goods has become the de facto mode of economic and social development (MacLaran and Kelly 2014a, 22). This has occurred in ways both imperceptible and disruptive, from "provision of public housing, hospitals, motorways, schools, rail and water and waste infrastructure" (Hearne 2014, 158) to directly incentivized development of private technological infrastructures like data centers. Corporate governance has become normalized to the degree that semi-state organizations like NAMA and the IDA, designed to partner with and smooth out space for the private sector and particularly FDI companies, are foregrounded by the state as providers of prosperity, jobs, and development, providing political support for private industrial or telecommunications infrastructure like fiber optic cables and data centers.[9]

Commercially oriented semi-state corporations and institutions in Ireland have a unique history, wherein for-profit public schemes like the SFZ, the Dublin Docklands Development Authority (DDDA) (in operation from 1997 to 2016), the IDA, and companies like Coillte (see chapter 4) and Bord na Móna (Bresnihan and Brodie 2025) demonstrate the semi-state administration of territory as experiments in regional planning and development. Closely partnered with multinational companies, they exist partially autonomous from the presiding government. So, while there is a justifiable tendency to chastise the private sector and appeal to the public for a more equitable and balanced system, in neoliberal societies, particularly Ireland, the distinctions between these fields (one ostensibly designed and operated in pursuit of profit, the other to provide services and regulate the private sector) become very unclear once you begin to untangle the history and functional operation of infrastructural networks themselves (see Plantin and Punathambekar 2019). Just as "the State came through the pipes" with modern infrastructural systems (Trentmann 2009, 303), so too does the corporation, and an overarching problematic of the field of infrastructure studies is studying these relationships and politics at the intersection of public utility and private ownership.

The public and the private are entangled across contemporary environments, and within our experiences of them, made to feel natural and occluding the material politics of these imbrications. Privatization thus requires a reconsideration of how infrastructural publics are formed, and how corporate entities, generally in partnership with the state, provide access to services that are infrastructural to everyday life (e.g., the internet). To media theorists Clemens Apprich and Ned Rossiter, the "spectre of the public" persists around anxieties of "access" (2016, 276), a privatized version of what public "goods" look like when telecommunications infrastructure is administered by corporate powers. Holt and Vonderau also signal toward this in their essay on cloud infrastructure, wherein politics and subjectivity are resituated within the realm of consumer goods but only strategically visible in such a way that these infrastructures appear apolitical (2015). But as these regimes are increasingly standard, we need to recalibrate politics to account for their infrastructural operations outside of the democratic purview of the "public," a sphere perhaps woven so tightly with the private as to become dangerously blurred through infrastructural provision and governance. Labor regimes and political subjectivity are fundamentally transformed across these environments. The market comes to take over spaces once reserved for politics as the place for public participation, citizenship, and care. Labor is no longer protected by state infrastructures, but is subject to the exploitation of private and transnational ones; politics reorients strategies toward private-sector inclusion, as the case of Athenry residents mobilizing politically for Apple's investment illustrates (see chapter 4).

Infrastructures and their politics are also profoundly affective, representing and containing the public aspirations of societies and the "promise" of modernization, development, and participation (Anand, Gupta, and Appel 2018; Larkin 2013). Media infrastructure in Ireland is a site of deeply affective politics, especially since the financial crisis and during the recovery, related to imaginations of progress, culture, public and civil participation, and conceptions of the country's social and economic futures. This is especially true with regard to global media and the tech industry, whose visual cultures—whether landscapes used to create *Star Wars* or tech's digital renderings of a clean and prosperous future—form an infrastructural allure in themselves. However, much as infrastructure is designed to recede into the background, it is within its invisiblized underbelly that the uneven mechanisms of its development and operation play out.

Political programs struggle to articulate these connections between the public and future-facing visible world of development and the actual material processes by which it occurs. Those "left out" of its development or stuck living off of the "ripple effects" of its construction and operation, as has been the case with FDI-led growth in the country, feel as though they are on the outside of these narratives of progress, leading to infrastructural dispositions of abandonment, stagnation, and the casualization of many forms of work across the country, especially in rural regions.[10] Geographers Ashley Carse and David Kneas (2019) and communication scholar Jenna Burrell (2020) argue in different contexts about the constituent *unfinishedness* of infrastructure for many subjects around the world, especially in rural and historically disconnected areas where projects tend to stall or never get off the ground, within which the state appears to them as antagonistic or even nonexistent (Chatterjee 2004; Gago 2017). Intimately related to privatization and financialization, in environments of crisis and indebtedness, how people relate to themselves, their communities, the state, and the market responds to particular conditions of what is materially available and how to make claims on it (Gago 2017). Whether expressed through entrepreneurialism (and other pro-business ideologies and activities) or political upheaval, populations are constantly involved in and interfacing between global processes, state programs, and more emplaced conditions of availability at the same time, even if their material temporalities are those of slowness and abandonment rather than a more spectacular kind of infrastructure promised by the "global."

From such examples we see that infrastructure and its operations do not recede into the background and become ambient and "atmospheric" (see McCormack 2018); they reproduce or provision consequential divisions of access and uneven development. In Athenry, where Apple tried to build an €850 million data center but had not expected a half-decade political fight through both the town's and the country's court systems, it could not recede into the background—its politics were too eventful, out in the open.[11] In west Kerry, bottlenecks and undeveloped roads mean that it is de facto cut off, by both landscape and remoteness. The same goes for any location where the basic rights of housing, health care, and other former "public" services fail daily.

Take basic internet telecommunication. High-speed internet remains unevenly available, and the state contracts fiber-optic infrastructure to private companies such as eircom, a formerly state-owned and now privately operated telecommunications company incorporated in Jersey. eircom, under the trading name open eir, was contracted as of 2020 to provide 80 percent of the country with broadband, whereas another company, National Broadband Ireland, controversially received a €3 billion contract to connect the remaining 20 percent of "rural users." Advertised as a "turning point for the revitalisation of rural Ireland" (Burke 2019), this expensive and clunky project was rolled out as broadband stood to be largely replaced by 5G technologies within the next few years. The public/private purview of public services development has caused a variety of problems, as these costly projects leave the state at the behest of companies focused on diversifying profits and public visibility, whereas the state, politically, seeks to provide equitable access to infrastructure across a nation still divided along urban and rural lines. Poor planning and messy rollout are bad optics for the state, leading to further public alignment with the private sector to manage and provide goods more "efficiently" due to access to capital and expertise. In this way, strategic development of media infrastructure in Ireland is a reminder of the variegation of power and sovereignty in contemporary life, or what Ong has called "graduated" or "variegated" sovereignties (2006), depending on the spatial and economic mechanisms employed. When critical infrastructure is designated to the private sector, and the environment itself becomes increasingly privatized, we need different ways of politically understanding the places in which we live, work, and struggle.

In my ongoing research into Ireland's infrastructural politics, I see frequently how communities and populations relate to these large-scale processes and institutions, both affectively and tactically. Citizens and workers often feel like the state simply cannot or will not help, its infrastructures not designed for their direct benefit, thus compelling them to withdraw from official state processes and engage directly with the politics of private-sector inclusion and expansion.[12] In each of the next three chapters, these spatial and infrastructural politics are unraveled in relation to their localized effects. In chapter 2, as the policy- and FDI-led "creative city" ostensibly thrived throughout the mid- to late 2010s, we will also see how alternative practices of reuse and spatial reclamation by and for residents were foreclosed by the state and capital. In chapter 3, we will unpack the media policy which is *infrastructural* to the operations of global media capital as it circulates through the country, which distributes effects unevenly and leaves workers toiling at precarious gigs in the dearth of direct public funding and subsidies. And chapter 4 will unravel how civil society mobilizes in sometimes populist ways for the private-sector infrastructure projects of big tech, in the case of Athenry, appealing directly to a multinational tech company while the state's plans foundered in the courts. In each case, politics are organized in and through the particular *mode of production* employed through *circulatory* infrastructures. We can thus observe how infrastructural politics, in Ireland and elsewhere, are mobilized by the "circulation struggles" (Clover 2016) of workers and communities that necessarily arise from sites of capital movement—whether via finance, media, or technology.

But infrastructure is not only about circulation. After all, industrial infrastructure has long been a way to *connect* production or extraction to circulatory apparatuses, whether via storage or transport—think grain elevators (Barney 2011), mining infrastructures (Meade 2017), or, in Ireland, railways built by Bord na Móna to transport peat (Bresnihan and Brodie 2023). In media infrastructure, then, we must analyze circulatory logic of technology and infrastructure supply chains to account for the term's usage within infrastructures like film studios (Dickinson 2024; Jacobson 2015), VFX hubs and facilities (Acland 2018; Chung 2012), and other sites of direct production for film and media. These sites facilitate the circulation of people, goods, information, and finance through Ireland. As Denis O'Hearn argued presciently at the height of the Celtic Tiger, the fundamental contradictions between Irish state development strategy and the labor and infrastructure required to support these investments would lead to significant bottlenecks, during which the state would either be forced to loosen constraints on public spending or face a reality where they could no longer provide infrastructure at the scale required by multinational companies locating in the country (2000, 88). The provision of public infrastructure, whether funded by the state or through PPPs, is a necessary basis for growth through FDI. But in the case of something like data centers (see chapter 4), which are extremely resource-intensive and entirely in service of private corporate profit through cloud and other services, this growth presents fundamental frictions with existing scales and rates of available infrastructural provision in an environment of constrained public spending.

We must also begin to include state infrastructure—such as roads, railways, airports—into the assemblages necessary to understand the environment of media and its production and circulation (and its possibilities for growth). These existing infrastructural routes help determine location and viability for industrial infrastructures. SI and other government bodies advertise public infrastructure as opportunities for companies to locate and perform productive activities, but the ways in which these activities generate value perhaps derive more centrally from efficient and interconnected modes of circulation. Supply-chain capitalism demands that productive activities must be coordinated across vaster spaces to innovate novel ways of cutting costs and incorporating receptive territories (Tsing 2009). The more effective the coordination, the greater the value produced (see Cowen 2014; Rossiter 2016). Whether the efficient shipping of filming materials in order to avail of otherworldly landscapes and largely nonunion labor crews in west Kerry, the logistics needed to transport talent to one of the film studios outside of urban Dublin, or the fiber-optic networks required to perform and circulate postproduction and effects work, these infrastructures are productive apparatuses that contribute to better managing and facilitating the flow of people, goods, information, and capital through Irish space.

None of this is to deny the specificity of media infrastructure, nor to reproduce the fallacy that digital media and supply-chain capital always interoperate smoothly.[13] Roads only become relevant to, and *as*, media infrastructure once they enter an assemblage of elements required for media capital to circulate. For example, deciding to shoot a TV series at Troy Studios in Limerick in 2017—near its inauguration and before a local workforce could be counted on—meant that equipment, crew, and talent would have to be initially transported from elsewhere, at which point, airports, railways, and other transport sites also enter into this network, along with pipes, cables, electricity, and the like. A Chinese tech multinational deciding to contract its data center space to a company in Arklow (in southern County Wicklow) also requires the effective operation and coordination of business strategies from central headquarters to European headquarters, the supply of building materials and computer components, the contracting of energy or electricity supply (preferably from renewable providers), the transformation of that supply into usable energy, and the functional circulation of data through privately owned and operated undersea fiber-optic cables—not to mention global users making and watching TikTok videos. The interconnectedness of labor and environments across all these points in media's supply chains helps us understand how the industry operates, including the contributions of transport workers, craft services, catering, builders, engineers, maintenance professionals, and security, "above" and "below-the-line." Infrastructural systems, far from operating in the background without friction or management, require tremendous amounts of labor and resources. If, as Curtin notes, media globalization has created a "global infrastructure of media labour exploitation" (2016), then we also make certain that we are attentive to the fact that supply-chain capitalism, in no way a smoothly integrated system, relies on patchy, variegated networks and spaces of subcontracting and outsourcing.

These imbrications become a geo-economic process by which the landscape of a given territory is produced, transnationally, as a space through which value can be extracted from an already-existing tangle of elements, such as a nation's internet infrastructure, road networks, or energy grid capacity. In data centers in the 2010s, this came largely from consumer video streaming and enterprise data mining, where extraction withdrew from the apparent abundance of data as a "raw resource" (see Apprich and Rossiter 2016).[14] Mezzadra and Neilson outline the reliance of the global economy on the nexus of "extraction, logistics, and finance," also as a frontier for the social production of value from flexibilized and distributed forms of production (2013b, 2015, 2019). The extractions *facilitated* by infrastructural logics, of course, have their own environmental consequences: Transport mobility for media production, boats laying cables, and data centers all share a necessity under current energy systems to burn carbon. Scholars have noticed that within supposedly "green" and technologized global supply chains, extractive activities have expanded rather than contracted with increased efficiency, whether in resource frontiers in the Global South for lithium batteries and computer components (Arboleda 2020) or through intensifying value extraction through consumer and enterprise technology (Srnicek 2016). Data centers, as a particular and media-oriented instantiation of this much wider system, like mushrooms popping up from miles of underground mycelium, allow us a chance to see where a spectacular version of the spatial development of "logistical media" undergoes transformations and friction via localized infrastructural politics. The "extractive frontiers" of data and smart devices are infrastructural to how we live our everyday lives, how we are governed, and how the future is unfolding (Bresnihan and Brodie 2021; Rossiter 2016).

Unpacking digital media's entanglements with communities, workers, and the "natural" environment, both discursively and concretely, Ireland's media infrastructural politics reveals a series of logics by which "extraction" of value from space and labor occurs *through* infrastructures. This is pertinent to the similar but specific disruptions and issues faced by tech infrastructure development in different parts of the world by large corporations. These disruptions, as will be described below, are the result of both global systemic interdependence and, increasingly, the turbulence of planetary ecologies. Each of these forms of turbulence is managed by logistical ways of thinking, and is shaping Ireland's media infrastructures and environmental futures.

Logistical Media and Supply Chain Extraction

The mechanization and optimization of logistics, along with the dispersal of production across global supply chains described above, has disempowered labor organization worldwide and increased structural precarity for workers and the social mechanisms in place to support them (Chua et al. 2018; Cowen 2014; LeCavalier 2016; Tsing 2009). The strategies and technologies through which this has occurred are increasingly technologized, automated, and naturalized (Levinson 2016; Mezzadra and Neilson 2019; Rossiter 2016). Although I aim to build on this kind of work, this book looks more specifically at how *logics* of capital organize certain commonsense ways of governing and managing labor, production, and circulation across global media infrastructures and supply chains during Ireland's decade of post–financial crisis austerity. Logistical rationalities have greatly intensified the relations of capital in and through the media infrastructural landscape of recovery-era Ireland, a developmental landscape through which multinational capital was able to innovate novel forms of extraction.

In this way, I am interested in logistics as a "calculative rationality" of capital (Chua et al. 2018)—a managerial exercise of sovereignty across logistical networks and infrastructures—that shapes practices and policies within the zones and territories through which corporations operate. Geographers Charmaine Chua, Martin Danyluk, Deborah Cowen, and Laleh Khalili account for this broader reasoning in a programmatic piece for a "critical engagement with logistics": "a wide range of circulatory processes—flows of goods, services, bodies, information, and capital—can productively be examined through a logistical lens. Viewed from this perspective, logistics is not reducible to a mundane science of cargo movement or a discrete industry among others. Rather . . . it is better understood as a *calculative rationality* and a suite of spatial practices aimed at facilitating circulation—including, in its mainstream incarnations, the circulatory imperatives of capital and war" (2018, 618, my emphasis). In Ireland, the suite of practices that facilitates circulation includes state policy, spatial planning, visual culture, and technological infrastructures, as much as more traditional logistical movements like the movement of raw goods and resources. As Danyluk (2018) and others have argued elsewhere, the financialization of the world economy has operated in tandem with the "logistics revolution" and its spatial and structural impacts on industries and infrastructures across the world. For example, the redevelopment of the Dublin Docklands, akin to other waterfront industrial spaces through the late twentieth century, followed infrastructural changes and the decline of employment in the Dublin port due to containerization, which moved operations away from the city center and to deeper waters closer to the mouth of the River Liffey at Dublin Bay. As later chapters will argue, global logistical infrastructure has implications far beyond only the material circulation of goods, as the rationalities of supply-chain management and security enter into the media and creative industries and their spatial practices.

Thus, while a materially specific managerial sensibility, logistics provides a lens to understand the material processes by which value is created *by* coordinating movement and the various strategies and tactics within (Bernes 2013; Harney and Moten 2013). This movement does not just encompass raw logistics—for example, the moving of people and goods on ships, trucks, and planes, across seas, roads, and skies. It is also about the circulation of media and information. Logistical technologies connect disparate places across the world, collapsing space and time, as the familiar description goes, and in the process eliminating or disempowering the role of labor. In doing so, the production of value is increasingly organized across these disparate supply chains, as the "circulation of stuff" means that commodities are manufactured and coordinated "*across logistics space*" rather than in singular locales (Cowen 2014, 2).

This dispersed landscape also, however, opens spatial and temporal fissures. Between the "logistical worlds" that are "stitched together" across the capitalist world system (Neilson and Rossiter 2017, 13), there are also alternative, and turbulent, "worlds." For example, in Ireland, the "wild" past activated by current business and tourism discourses often proves an unruly opponent to the smooth flow of capital upon capital actually hitting the ground, whether the choppy seas off of west Kerry stopping boats from reaching a *Star Wars* production location or the resistant forms of social organization that have disrupted pipelines and cables from Shell, Facebook, and Google in County Mayo. In this sense, logistics and its spatial technologies are always navigating a lumpy past and a turbulent present to conjure a future world for capital. "Logistical worlds," then, even in the imagination of capital, are not fixed—they are always in motion, and in so being, logistical rationalities are constantly encountering (and suppressing) unruly spaces and subjects. Logistics and turbulence coevolve together: As turbulence yields new contingencies for logistical rationalities to manage, logistics produces new turbulence.

Securing the production and extraction of value across space is thus an infrastructural process of managing, stamping out, or stitching together the unruly "worlds" of circulatory capitalism, whether discussing finance or logistics. Mezzadra and Neilson argue that an integrated approach to "extraction, logistics, and finance" is necessary to "pursue an analysis that underscores the crucial relevance of capitalist activities within specific economic 'sectors' without succumbing to arguments that position such sectoral activities as exclusive frames for the interpretation and contestation of contemporary capitalism and neoliberalism at large" (2017, 187). Tsing performs a similarly interlocking analysis, arguing that while "supply chains are not the only forms of contemporary capitalism," which include finance and large corporate entities, they exemplify the simultaneous heterogeneity and standardization that characterize the diverse operations of the world economy, within which "diversity forms a part of the structure of capitalism rather than an inessential appendage" (2009, 150). These operational logics that Mezzadra and Neilson and Tsing trace across apparently loosely interconnected sectors of the economy are central to the extractive and exploitative activities that occur across their specific sites and centers of power. In this way, the financial extraction of debt by the IMF from countries in the Global South can be linked to the developmental activities of the PRC building ports and roads—logistical infrastructures—for economic development;[15] or the circulation, accumulation, and extraction of value from data in mechanical sheds encircling Dublin can be intertwined with the financial apparatuses that funnel wealth through tax evasion back to headquarters in California; or how the Irish state *provides* the institutional and infrastructural measures for multinational tech capital that connect the country with the mining of rare earth metals and the exploitative manufacturing of computer components in central Africa and China. Logistical worlds, stitched together by *logics and technologies of accumulation*, for better or worse.

But as many have pointed out (see Cowen 2014; Mitchell 2011; Tsing 2009), despite more intensive technologies of controlling workers and sophisticated modes of operating across diverse territories, the extensive character of logistical networks makes them subject to more disruption across their dispersed activities, including the potential "counterlogistics" of labor struggles (Bernes 2013) and the "unmanageable" contingencies of broader geopolitical disruption. Chua, Danyluk, Cowen, and Khalili propose that in studying logistics, we must account for turbulence, but also for the frictions of spaces through which such logistical flows travel, the "lively" social worlds that logistical infrastructures encounter and attempt to "co-opt, contain, or absorb" (2018, 623). As these authors argue, "the gap between the idealized imagination of logistics and its messy implementation reveals that the project of making the world safe for circulation is always incomplete" (624).

Bridging these gaps is one of the many hurdles for a logistical vision of the world, and where media studies is uniquely poised to understand logistics.[16] Chapter 2 asserts that the visualizations of the city made by digital media companies for developers demonstrate a *logistical* process of preparing the city for financial circulation, but cannot contain the unruly and chaotic spaces of the city. So-called "fast-track planning," which attempts to remove local dissent from the planning process, nonetheless encounters resistant political formations. Chapter 3, then, builds on spatial understandings of global media supply chains by looking at how the state's media policy conditions Irish space for FDI and offshored media production, in doing so failing (or choosing not) to provide care for the country's "labor resources." Circulatory capital must always manage turbulence—or liveliness—at a variety of scales, as turbulence is a condition of life across both the systemic and the everyday. For Tsing, "Supply chains are harder to control than corporations or state bureaucracies," as they are forged "within the constant flux of boom and bust opportunities" (2009, 150). Supply-chain disruptions span "the everyday delays of bad weather, flat tires, failed engines, missed connections, traffic jams, and road closures" (Cowen 2014, 2), and even, as chapter 4 will detail, the growing turbulence brought by climate change.

A variety of methods and machineries are employed to manage the "turbulent" and "lively" worlds confronted in logistical operations, and such technologies are a large part of the logistical apparatus. What Rossiter calls "logistical media" (2016), after John Durham Peters (2015), describes the "technologies, infrastructure, and software" that "coordinate[s], capture[s], and control[s] the movement of people, finance, and things" (Rossiter 2016, 5). In Peters's and Rossiter's conceptions, this media is often *atmospheric*, in that it operates across a political ecology that includes technologies, elements, minerals, and people in its very makeup. The expansion of global communications infrastructure culminates in the governance of geo-economic trade and financial flows facilitated by a variegated and highly privatized network of communication technologies, a system of fiber-optic cables, data centers, satellites, office spaces, and the hardware and software meant to ensure smooth operation and minimal interference. This contributes to the increasingly naturalized operation of these companies within economic and political life. How does one discern the difference between an Amazon manager and a state official effectively *telling* you that they are going to, not asking you if they can, build a data center somewhere? The lines become blurred. As these infrastructural environments appear ambient and circulated by state and corporate power in shared measure, it grows more difficult to govern by the state's terms (and easier to govern in logistical terms).

But contrary to perhaps pre-emptive statements about the sovereignty of these companies *over* spaces, wherein chameleonic multinationals like Amazon are seen to crunch increasingly dispersed aspects of social, cultural, political, and environmental life into their operations, we must pay close attention to exactly how political subjectivity and governance are reshaped through their interactions with places and nation-states. US multinational tech corporations have maintained a presence in Irish public life for long enough to have permeated the social and political fabric of the country, with Apple locating in Ireland in the 1980s and IBM as early as the 1950s. While many of these companies are much newer, like Amazon and Google, we still cannot think of their individual "incursions" as somehow fully external to the Irish state, parachuting in unannounced; rather, the myriad partnerships and negotiations built by the tech and finance industries more broadly with the Irish state means that their presence has become somehow endemic to how the state operates, a process that was innovated, I argue, through the financial crisis and recovery in the 2010s.

If finance organizes and logistics manages natural turbulence in Ireland, they do so through the transnational corporations operating through it and in partnership with the state. So, while Ireland might not be a dystopian "logistical state" as theorized by Rossiter (2016), it remains true that the expertise, training programs, and longer-term contracts that multinational companies have with the state position them as service providers in an increasingly privatized environment. These "ripple effects" are seen as primary benefits of FDI-driven operations to local communities (see chapters 3 and 4). These infrastructures only "become sovereign," Rossiter argues (2016, 5), after a series of complex negotiations with the state leaves people and workers subject to the coercive power of transnational business at high levels of planning and development. This corporate sovereignty is then felt *within* the everyday struggle of making a living in a place where the benefits of big capital are experienced, at best, in trickles rather than waves for all but a privileged few.

Digital technologies coordinating labor and environments represent the infrastructuring multinational tech through "smart" environments and spatial planning (see Gabrys 2016; Mattern 2017), as well as the sensors and datafied technologies that will optimize the generation and delivery of electricity via the atmospheric "resources" of wind and other renewable energy (see Bresnihan and Brodie 2021). While atmospheric in their own ways, financial, logistical, and governing systems traditionally have a hard time dealing with the unpredictable planetary turbulences of meteorological and geological systems, and digital technologies are employed to manage these uncertainties. For example, financial systems must increasingly account for and predict climate turbulences (Cooper 2010); undersea fiber-optic cables face disruption from both fishing activities and climate change (see Starosielski 2015, 2019); and even the basic chore of getting people and materials to filming locations in Ireland can be disrupted by bad weather, as the treacherous journey out to Skellig Michael in west Kerry proves for both global film production and tourism, which will be analyzed in chapter 3. At all scales, from protection to insurance to snow tires, financial and supply-chain security thus must account for classic as well as emergent disruptions, requiring flexible, often real-time digital capture and analysis via sensing, mapping, monitoring, and other datafied solutions even in diverse, dispersed, and often mundane contexts. However, as infrastructural networks providing public and cultural goods are subjected to greater turbulence, so are the administration and governance of these goods, data, and people beholden to these logics of management that attempt to not only control, but to extract value from such unruly conditions through infrastructure.

Data centers and other big tech infrastructures are not only sites through which extractive activities of logistical media *happen*, they are extractive technologies in and of themselves. Just like spaces where media production occurs throughout Ireland, if extraction across capital's frontiers is intensifying through the logistical organization of its sites of value production, then we can observe that processes of extraction occur not just *within* logistical networks but *through* them. "Calculative rationalities," like financialization, are not simply learned in schools, deployed in boardrooms, and felt by specific workers. They are collective and agential systems of thought through which economies function and power operates. Extractive processes extend into our very ways of seeing and moving through the world.

Conclusion

If this whistle-stop tour from the early, postcolonial Irish state's industrialization plans to the current juncture of multinational tech-led changes in investment and governance in the post-crisis Irish state seems a bit dizzying, perhaps we can take a moment to reflect on the accelerated experience of these developments *within* a globalizing Ireland and the sense of whiplash that the country must have felt through the latter half of the twentieth century. Turbulence was not limited, after all, to the global economy. Places across the landscape experienced more troughs than peaks in this process, with minimal multinational liquidity nourishing deprived spaces across rural Ireland and the notoriously poor Irish cities from the 1960s to the 1980s, at least for those most marginalized.[17]

Nonetheless, we must also be careful not to associate Ireland with only its reactions and experiences of large-scale capital, globalization, and state policies. As anthropologist and critical theorist Elizabeth Povinelli has so meaningfully articulated, the "eventfulness" of the crises of what she calls late liberalism masks the underlying, uneventful, and everyday experiences of just getting by for most of the world (2011). While this chapter acts at the macro scale, as an infrastructure scholar, I have for the rest of the book resisted the urge to *make eventful* the spectacular projects of state and capital's liberalization and industrialization, attuning to the slow, protracted, and quotidian experiences of infrastructure as "unbuilt and unfinished" (see Carse and Kneas 2019) for many, especially in the way that the process of *making infrastructural* happens at ground level (see Ruiz 2021). This view is available to us even if we remain dialectical in our understanding of how infrastructural power is developed and administered by hugely powerful formations.

The diverse and extractive operations of contemporary capital, then, require site-specific and rigorous empirical work, but they also demand transdisciplinary grounding and a systemic understanding of global capital's territorial wages. In bringing together concepts and methods from anthropology, critical geography, media industry studies, and science and technology studies, particularly through the history of financialization, the globalization of media industries, and the role of media infrastructure and so-called "logistical media," I have in this chapter begun to unravel the entanglements that occur at sites through which capital is operationalized in post–financial crisis Ireland. We can now see how the *making infrastructural* of tech and media capital occurs in part through a series of ongoing and pervasive naturalizations of the role in capital in bringing and sustaining economic and spatial development, especially since the 2007–8 financial crisis, during which the extractive operations of FDI-driven investment intensified through an environment of austerity in the 2010s. In the coming chapters, I deepen this analysis through engagement with the country's policy and planning environment and recourse to a variety of sited case studies across Ireland.

In bringing together these global concepts, sites, and literatures and navigating through their complexities via the Irish context, I am emphasizing the ways in which certain rationalities of spatial development, policy, and management have transformed and come to appear as necessary and *infrastructural* to the environments through which they operate in a way that extends far beyond the country of Ireland. Largely US-based FDI companies, many in the technology industries, have colonized the provision of public life throughout the world, from culture to politics to the very storage and circulation of our personal and embodied data. So, even while data centers and so-called "logistical media" are relatively novel technological infrastructures dictated by increasingly entrenched "calculative rationalities" governing space, in Ireland, as this chapter has demonstrated, they also represent relatively coherent—if still turbulent—progressions from earlier forms of economic extraction, privatization, and the management of turbulence that developed throughout Irish (neo)liberalization. Ireland, in the "recovery era," as the remainder of the book will focus on, in enabling novel and innovative ways for multinational capital to navigate and profit from crisis, acted as a kind of laboratory for emerging forms of value extraction—from what I call a "creative fix" in the ruins of finance capital, to experiments in extracting value through regionally distributed film policy, to the development of "smart" infrastructures like data centers. At the same time, residents, workers, and activists in Ireland found similarly inventive ways to weather and resist these turbulent innovations.

2 —— Ghostly Currents and Creative Erosion

It is winter 2019. On the south bank of the River Liffey, in a corner of Ringsend in sight of the International Financial Services Center (IFSC) and the rest of Dublin's high-finance North Dock, there is a boat in a car parking spot. It is out front of a residential apartment building, resting elevated near the mouth of the River Dodder and the Grand Canal where they meet the Liffey (figure 2.1). This boat, brought to land by some unknown tide, at first glance seems forgotten, a ruin of another time. Maybe of a time when residents in this portside neighborhood were more traditionally seafaring— although this is unlikely, as the boat's structure looks more or less contemporary, a speedboat or small fishing vessel. Perhaps, then, it was of a time when unexpected wealth drove some local resident to purchase the boat for leisure, which became impossible to maintain as whatever tide keeping the boat and its owner afloat receded.

During my fieldwork, when I would visit this area to photograph the progress and visual culture of nearby development projects, I was fascinated and confounded by this boat. Because it was not entirely forgotten. It was being labored over by *someone*, dismantled, neatened, or stripped for parts, either by its owner, local residents, or enterprising passers-by (figure 2.2). As an outsider,

2.1 A stranded boat, with Capital Dock in the background, in June 2018. Photo by Patrick Brodie.

2.2 The same scene in February 2019. Photo by Patrick Brodie.

I never learned the boat's full story. How had it arrived? Did residents consider it an eyesore? A depressing remnant of another time? Or did they cease to notice it, losing what might have brought it there in the first place to local history? What would become of it?

As I returned continually to Ringsend from 2017 to 2019, this boat became a striking visual and spatial metaphor for the financial crisis and its recovery. While the boat was progressively ruined and dismantled, a diverging future took shape directly across the canal—Capital Dock, a high-rise, mixed-use multinational investment property, rose higher and higher. At the edge of the "Silicon Docks," Capital Dock and its like uphauled space to create new futures out of former industrial and financial ruin in these former port lands.

I never learned the origin nor the fate of the boat, partially because of my insufficiently anthropological engagement with the surrounding area. My approach was always more topological than topographical. After all, as geographers Cian O'Callaghan, Sinéad Kelly, Mark Boyle, and Rob Kitchin argue, there were both "topologies" and "topographies" to Ireland's financial crisis—the former available to uncover hidden systemic processes and meanings, the latter to excavate deeper local histories and cultural meanings (2015). These processes and histories lived in physical spaces and practices, in vital relation to the surrounding environment, uncontainable and finding unexpected relations and reuse. Throughout my research, I encountered many such sites of informal and unfinished urban artifacts. Building on these mostly topological experiences of the city, this chapter unpacks the cultural, visual, and affective economies circulating through the built environment of Dublin as a "creative city." While O'Callaghan and colleagues unravel the varied histories and spatial signatures of neoliberalism across the country, in this chapter, I will focus on the role of media and visual culture, and their industrial and affective infrastructures, within the form and experience of the post-crisis creative city in Dublin.

The Dublin Docklands will serve as the center of this analysis, former industrial and port hinterlands in the heart of Dublin redeveloped as a cultural and enterprise district. These "Silicon Docks," in reference to the big tech hub of Silicon Valley in California, were the invented center of big tech capital in Ireland after the financial crisis. Google, Facebook, and Airbnb, among other companies, maintained headquarters in this area, attracted by Ireland's pro-business environment and low corporate tax rate (12.5 percent at the time), as discussed in chapter 1. Initially put on the foreign direct investment (FDI) map as the site of the IFSC, established in the north docks in

1987, the area now also houses some of the country's largest cultural infra-structures, like the Bord Gáis Energy Theatre. Administered by the Dublin Docklands Development Authority (DDDA) until 2012, and designated a strategic development zone (SDZ) for regeneration through the financial crisis, the National Asset Management Agency (NAMA), the Irish national "bad bank" set up in 2009 to absorb toxic real estate loans following the financial crisis (see Byrne 2016), centered much of its Dublin focus on this and the surrounding area for profitable commercial redevelopment using the SDZ mechanism (National Asset Management Agency n.d.). Repre-senting an increasingly familiar transition from financial to tech services FDI as a motor for the Irish economy (Brodie 2021), the location of US tech multinationals across the river from an ur-site of Ireland's financialization was long in development.

In the Docklands, like former port and waterfront spaces in Bilbao, London, and many other places globally, what used to be port and industrial hinterlands fell into ruin and disuse through the late twentieth century, only to be "rejuvenated" as so-called "cultural quarters" by cultural infrastruc-tures and creative enterprise in the twenty-first century (see, for example, Simpson 2017, 132–36). These places were remade in postindustrial society as zones for cultural experience, innovation, spectacular wealth, and con-spicuous consumption. The logics of development that played out across the Docklands demonstrate the structural role that "culture" and "creativ-ity" played in the redevelopment of Dublin through crisis and recovery from 2007 to 2020. While industrial structures outside of the central areas of this district decay, awaiting a new future, inside the creative city, this futurity was crafted through multinational investment with spectacular abandon. At the time of my fieldwork between 2017 and 2019, few visible ruins remained that were not soon eaten up by the creative city's circulatory machinery, as billboards, construction hoardings, cranes, trucks, artworks, and graffiti coexisted in the same spaces, on the same streets, in the same alleys and public squares, forming an overwhelming and choppy experience of the city and its real and imagined ways of life. This chapter argues that this represented the redirection of capital, led by austerity policy, through neoliberal regimes of culture and "creativity."

During my fieldwork, you could pick out the Docklands by the sea of cranes rising into the sky.[1] The area borders several working-class residen-tial neighborhoods, such as Irishtown and Ringsend to the east, the latter home to the peculiar boat described above. Companies like Google, one of the several US tech multinationals headquartered in the area, have tried and often succeeded in buying up property in these neighborhoods, speculating on the expansion of the Docklands' spatial influence. On the other side of Ringsend and Irishtown, the Poolbeg Peninsula extends outward into the sea. The Poolbeg West SDZ, consisting largely of vacant and derelict industrial property, was earmarked in 2016 for mixed-use properties by multinational developers, who fought with a high-profile film studio project called Dublin Bay Studios to obtain permission for the site from the Dublin Port Company (DPC). At the far eastern end of the Poolbeg Peninsula is the Pigeon House Power Station and Hotel, at the time of writing largely abandoned industrial structures neighboring a controversial waste-to-energy plant, owned and operated by US-based waste multinational Covanta, and an Electricity Supply Board (ESB, a state organization) power station. The Pigeon House site was the subject of debate as to the location of a "cultural precinct" and "creative cluster" nestled within the loud, polluted industrial zone (Kelly 2018), and has acted as an ad hoc filming location for media projects including fantasy flop *Reign of Fire* (Rob Bowman, 2002).[2] Land availability and the productive forces of media capital, incentivized by industrial development policy, overlapped in speculative plans for the peninsula, eventually underwriting its fate as a mixed-use extension of the Docklands.

This chapter will examine these ruins, speculations, and paths not taken surrounding the spatial development of Irish media industries in the post–financial crisis era. Media's role within the post–financial crisis city represents a series of encounters that provide a glimpse into the wider operations of capital in conditions of recession. This is true whether looking at the common practice of renting out ruined industrial spaces and warehouses for global media production due to the *absence* of formal studio infrastructures; the branding of "heritage" to advertise and put a futuristic veneer on disused buildings; or bidding wars to activate formerly industrial and port lands for creative city projects. Focusing on the role of media in the construction and imagination of built space, specifically through the lens of financialization, property markets, and spatial development strategy, this chapter also attends to the affective dimensions of the "creative city" and its differential distribution of cultural celebration and belonging. The creative city was never designed to be an inclusive space, and in fact its very design (enacted through spatial development) ensures that its urban playgrounds become privatized and placeless utopias for well-off consumers. The collapse of culture and profit in the spatial form of these strategies, which proliferated across the world through the 2000s but especially post–financial crisis, positioned tech and media sectors to uniquely intersect in the form and function of these financialized zones. Using culture and creativity as Trojan horses to promote the "public good" of these sorts of projects during austerity, the state allied with financial power in effectively bulldozing any alternative visions of reuse that might arise in the ruins of finance capital. In doing so, it shifted the violent turbulence of the market onto its own resident "public," which coexisted uneasily with the more stable and fixed visions of capital's imagined future. This chapter locates and traces the contours of a variety of projects centered around the media industries specifically within post–financial crisis Dublin to better understand how finance capital operates *through* the spatial conduits of media and creativity. To conclude, I show how art and culture that behave in truly fugitive and antagonistic relationships to these manufactured spaces, living with rather than against "ruin," point toward foreclosed political opportunities presented by the crisis.

Ghost Estates, Property, and the Cultural Mediation of Crisis

At few points in recent memory was the residual turbulence of global capital visualized more vividly than by what came to be known in Ireland as the "ghost estate phenomenon." After the 2007–8 financial crisis, built environments across the country were left in stages of incompletion and disrepair after the abrupt recession of capital from the property and construction sectors. So-called "ghost estates" dotted the landscape, largely suburban and mixed-use properties stranded by the retreating tide in various states of half-construction. Ireland's landscape was—and in some places still is—dotted with these reminders of the crisis, built by national and international developers lured by property tax incentives and fueled by the bad loan practices of both domestic and foreign lenders. When the markets crashed worldwide, as chapter 1 narrates, the materiality and interdependence of global financial systems were laid bare, as capital receded, construction halted, and those responsible abandoned their investments and were bailed out by the state, leaving these structures behind. Concrete, rubble, overgrowth, disconnected wires and uninstalled pipes, abandoned construction equipment, and skeletal structures characterized these incomplete sites, usually housing estates or mixed-use development projects, ironically mirroring, in a more stagnant way, the rubble-infused chaos of the creative city in Dublin. But the most profound measure of their significance was the eerie, resonant feelings of guilt that haunted them, the spectral presence of foreclosed futures, washed away by the changing tides of an unstable and imbalanced global market (figures 2.3 and 2.4).

2.3 Photo of ghost estate in rural Ireland by Valérie Anex (2011). Reproduced with permission of the artist.

2.4 Photo of a ghost estate by Valérie Anex, with foreclosed suburban future visualized (2011). Reproduced with permission of the artist.

These abandoned and unfinished spaces were both pregnant with and emptied of the "excess" of the Celtic Tiger, and came to symbolize the "ruin" of Tiger-era prosperity and the punished "excess" of the Irish populace (see O'Callaghan, Boyle, and Kitchin 2014). There was significant reflection, in both fictionalized and nonfiction artworks, accounts, and essays, about their status as a visual representation of the evacuation of Celtic Tiger promise (Anex 2011; Carswell 2017; Haughey 2018; McWilliams 2006; Teicher 2014; Wall 2011).[3] Photographer Valérie Anex's *Ghost Estates* photo series evocatively visualizes the eerie structures (see Teicher 2014; also Anex 2011), hauntingly documenting these abandoned environments against the stark backdrop of gray Irish skies. Anex refers to these rural and suburban sites as a "topology of the economic disintegration of the country" (Anex 2011), as the tides of remarkable prosperity receded even more rapidly than they had risen. Artistic reflection processed the feelings and foreclosures of such complex economic and affective entanglements, as this crisis of the economy became rapidly a crisis of national character and failure, especially for those most severely affected.

The emotional and affective dimensions of economic development and the 2007–8 financial crisis specifically have been widely discussed, both in and outside of Ireland. Political economist Maurizio Lazzarato memorably details the subjective and affective constitution of "indebted man" through the insidious spread of debt mechanisms (like those of the troika) and their forms of violence post–2007–8 (2012), echoing financialization's spread through culture and society as described in chapter 1. Cultural theorist Lauren Berlant argues that the "crisis ordinary" felt across the diverse worlds of neoliberal governance shapes affective responses to structural conditions and how they are mediated (2011, 10). Berlant's analysis traverses various scales of experience: "The ongoing present is also the zone of convergence of the economic and political activity we call 'structural,' insofar as it suffuses the ordinary with its normative demands for bodily and psychic organization" (17).

However, in Ireland, the "violence" of finance capital (Marazzi 2011), expressed through the punishing regime of debt, was frequently and publicly mediated through space and ruin, especially in the ghost estate phenomenon. As visible evidence of the finance industry's invasion of a symbol of domesticity—the house (see Richman-Kenneally and McDiarmid 2017)—the everyday lives of residents across the country were revealed to be entangled with the ups and downs of global finance traced in chapter 1, a form of value that permeated the landscape and social relations within it. The urban and suburban construction boom of the Tiger was likewise associated with imaginaries of class ascendance through simultaneous dynamics of global-ization and existing social stratifications in the country (see Kennedy 2003) and was built on the back of the "immaterial" and affective labor of social reproduction, actually and symbolically associated with middle-class and suburban ways of life and gendered divisions of labor (see Barton 2018). This is crucial to finance's particular mode of value production. By mediating its invisible circulations through the built environment and its financial instru-ments, it harvests surplus value from sites of production within circulation, through interceding structures like derivatives (LiPuma 2017) and in para-sitic material relation to industry, labor, and social reproduction. Like the industrial built environments before them that had fallen into dereliction with changing modes of production and social constitution, the decline of these sites, alongside and as part of the "boom building" that had become so pervasive and central to Irish life, opened profound questions as to the actual and future social costs of economic growth to the populace.

So-called "boom building," some unfinished remains of which still made up ghost estates across the country in the late 2010s, is central to ret-rospective narratives of the Celtic Tiger. These ostentatious developments and their ruins, associated with the "out-of-control" spending and irrespon-sible indebtedness of the Irish middle classes,[4] fed public condemnations of the "excess" of the Celtic Tiger. By this logic, the Fianna Fáil government's promotion of the narrative that "*we* all partied," "it's *our* fault," and thus could not be trusted with such prosperity, ultimately manufactured de-feated consent for austerity—in spite of the clear and multiple social strati-fications between those who reaped the success of the boom and those who suffered most in its aftermath (see Coulter 2015; Hearne 2013; O'Callaghan, Boyle, and Kitchin 2014).[5] As chapter 1 narrates, Ireland post–financial cri-sis was subject to the austerity programs of the troika, which imposed debt measures on the Irish state and populace, making the country dependent on their loans and draconian bailout conditions for economic sustenance.

Many of the troika's conditions were centered on privatization via FDI, which compounded the structural conditions behind the crisis in the first place. Irish popular discourse frequently heralded a kind of return of the colonial repressed, as subterranean currents of uneven development and dependence bubbled once again to the surface:

> One of the ways in which this shame and frustration was dealt with was by internalising the responsibility for the crisis. This position, some-

what ironically, offered a way to reconcile the nation's post-colonial history in a post-Celtic Tiger context. . . . This discourse implied that by stoically accepting responsibility Ireland could demonstrate its ability to stand on its own feet and avoid returning to a position of dependency. The supreme irony here is that by accepting the responsibility to bail out the banks, the Irish Government locked the Irish people into a structural adjustment programme that institutionalised dependency. (O'Callaghan, Boyle, and Kitchin 2014, 131)

The productive logics of finance, and its expansion into the everyday feeling of those living in and through its crisis, had to be filtered through public memory still rife with unresolved postcolonial dysphoria. As geographers Denis Linehan and Caroline Crowley note, the very "soul" of the country was opened up (2013, 4), as a crisis of feeling (pride, heritage, and independence) accompanied a crisis of the national economy (beholden to global finance). This cultural and affective dimension of the Irish financial crisis, a real and symbolic (over)presence, suggests that the 2007–8 crash became a crisis of culture as much as an overaccumulation of capital, symbolized within the ruinous built environment. The infrastructures and built environments that housed the overaccumulation of capital (O'Callaghan et al. 2015, 33), excessive in their own right, showed how quickly global capital could abandon spaces deemed no longer valuable, a spatial reminder of both what was and what might have been.

Within this environment, cultural life was frequently promoted as a site through which to repair the ruined sociopolitical landscape. As Linehan and Crowley note, in popular and official reappraisals of Irish identity following the highly emotive feelings of betrayal following the crisis, "it was culture, community, locality and creativity that were grasped as the authentic touchstones of who the Irish are" (2013, 4), as this crisis of the economy was expressed as a financialized crisis of feeling. Public aspirations of alternative futures and reuse were expressed through cultural means, negotiated through the public sphere. But these "shards of hope" (13) that culture and creativity offered were soon colonized by finance, as the question became not what sort of *repair* or *alternatives* culture could offer, but how it could be made *profitable* through a variety of spatial measures.

Logics of how to repair and care for these broken environments of finance capital took on a fascinating life,[6] even if quickly foreclosed by profit-driven measures of privatization and austerity. As O'Callaghan, Boyle, and Kitchin articulate, the media narrative and imagination of ghost estates were crucial elements in the popular negotiation of the crisis and the potential for future configurations out of rubble: "Within this narrative there were divergent agendas. For instance, community and sectoral advocacy groups suggested using 'ghost estates' for social housing or to accommodate growing homeless populations. Underpinning these proposals was a welfarist agenda to appropriate vacant properties away from the (failed) market to achieve a social dividend. Meanwhile, different groups put forward a series of 'creative' 'alternative' uses the estates might be put to, such as start-up or pop-up space for new businesses, film sets, or test-labs for smart city technologies" (2014, 127). They reference an article by journalist Gemma Tipton written in the immediate aftermath of the crisis, in which she invited architects to imagine unconventional and sustainable reuses for these sites, emphasizing the ways that the excessive futures of these spaces in the early stages of the crisis were full of hope for possible alternatives to FDI-driven development (2009). The answers ranged from innovation hubs for creative entrepreneurship to giving the spaces back to local communities and allowing them to self-determine their future.[7]

As anthropologist Nicholas D'Avella describes what he calls "concrete dreams" in Buenos Aires after the 2001 financial crisis in Argentina, thick and varied experiences of urban built spaces, especially in the midst of financial crisis, are not "concrete" in the conventional sense but rather always under quotidian construction and reconstitution based on local practices and needs (2019). Similar to the series of ground-level and community projects D'Avella describes, different ways of seeing properties and spaces in post–financial crisis Ireland seemed on the precipice of action, taking power and self-determination out of crisis back from the state and the "market." The website *Irish Environment* urged readers in 2012 to

> identify the nearest ghost estate to you and develop proposals for using those houses for some public benefit. Include in your analysis the cost to finish the houses, the legal ownership of the houses, the needs for houses in the area, and the means for raising funds to implement your plan. If one option being considered in your area is to tear down the houses, what use can be made of the site and who should control that decision—the developer, the local authorities, the banks or NAMA? What should be the role of the local community in such decisions? (Irish Environment 2012)

Such public solutions had gained traction, at least among some communities. However, one proposal, commented on by O'Callaghan, Boyle, and

Kitchin, was perhaps telling of the actual future that would emerge: Irish blogger Tom Dowling (a former employee at Troy Studios in Limerick) pitched for ghost estates to be repurposed as outdoor film sets for global cinematic production, as an alternative or competitor to Pinewood Studios in the UK (2011). Contained within such proposals, and similar to what Shannon Mattern identifies as the future-facing foil to repair—that is, innovation (2018)—is a commonsense logic that can be found across most economic development proposals in Ireland: To "resolve" stagnant assets, they must be brought back to life by new waves of foreign capital.

NAMA, an exceptional semi-state asset management agency put in place to resolve overaccumulation in the residential property sector, was designed to clear up these spatial fragments of finance capital without calling attention to its core operations. At a basic level, NAMA was designed to make indebted property profitable again: "Unlike most businesses which aim to expand over time, NAMA begins life with a very large balance sheet and aims to shrink it over time. It must do so while obtaining the best achievable return for the State on the assets it has acquired" (National Asset Management Agency n.d.). While these "returns" were meant to be directed for public benefit, there was widespread skepticism surrounding NAMA's redistribution of "toxicity" among a variety of smaller ventures and stakeholders. These ventures and stakeholders were largely developers and investors, who flipped these "assets" back on the market as property and rental costs exploded in the late 2010s.

To give an example: NAMA claimed in early 2017 that there were only twenty-five ghost estates left to be "resolved," down from 668 as of 2015, and that these remaining estates would be resolved by the end of the year (Hancock 2017). By January 2018, NAMA had apparently "cut exposure" on remaining ghost estates to eight (RTÉ 2018). However, it was unclear what NAMA actually meant—and did—when claiming these resolutions. Reports and figures always varied as to the actual number of total vacant units, and what these figures meant (O'Callaghan, Boyle, and Kitchin 2014, 128–29), if anything, which served to further obscure the "reality" of the crisis and its lasting implications. While the technical definition of a "ghost estate" was limited to "all properties built post-2005 where 10 or more units share the same estate/street address and more than 50 percent are coded as either vacant or under-construction" (Kitchin et al. 2010), other unfinished, ruined, and vacant built spaces also characterized the post–financial crisis landscape. Whether empty single-family housing developments in rural Donegal, never-occupied high-rises in Sligo town, the long-troubled

Pálás Cinema in Galway, unrented consumer spaces in the north docks, or aborted Luas (tram/light rail) stops through office parks in South Dublin, during various field visits from 2017 to 2019, these ruins remained, visible evidence of a crisis that many saw as "resolved." During the recovery, with the crisis technically "resolved," NAMA continued to operate. They still controlled stakes in various property development schemes and projects, for example, Capital Dock, Boland's Quay, and the Poolbeg West SDZ developments in the Docklands (detailed below). Renewed growth did not dis-embed crisis management from the DNA of Irish development—in fact, it crystallized it.

The remainder of the chapter will unpack the specific ways in which finance capital, the state, space, and media industries were entangled in Dublin's "creative" spatial development strategies in the 2010s. NAMA and its partners administered an ostensibly culture-driven spatial fix in Dublin through the continuing privatization of culture and space, siphoning capital into the "creative industries" and logics of the "creative city," and its attendant strategic zones, arts districts, and tourist spaces, which inevitably exploit, extract from, or displace existing communities, labor, resources, and infrastructures. The emotional currencies that culture carries with it—creativity, aspiration, heritage, community—were uniquely positioned to circulate capital through the built environment, and to build consent around the redirection of public resources into private projects and infrastructures by capitalizing on the affective and communal "dreams" of cultural activity. As urban sociologist Allen J. Scott argued in 2008 at the cusp of these global developments, "economy and culture appear to be converging together into new and peculiar structures of meaning whose focal points are the great city-regions of the global era" (vii). The explicit colonization of culture with "creativity" within the thought of planning has implications beyond more traditional media industries and production, as will be described in more detail below. They represent the violence of global capital implemented and weaponized toward—and by—strategically defined aspects of culture and labor.

Creative Labor, the Built Environment, and the Spatial Fix

The Dublin Docklands are a textbook example of the creative city model of governance. A designated SDZ on former industrial port lands, they housed at the time of my research European offices of tech multinationals Amazon, Facebook, Google, and Airbnb.[8] As chapter 1 details, the district was deregulated during the late 1980s and early 1990s to attract investment through the IFSC, where a corporation tax rate of just 10 percent on offshore profits was adopted in 1987. The district has undergone continuous "regeneration" and expansion, in particular through expanding mixed-use SDZs, where the state facilitates fast-track projects by private developers for residential, mixed, or purely industrial use. The creative industries (particularly high tech and media) were given special purchase in these urban plans over the preceding two decades, whatever the controversies over their measurable success and benefit to local residents. As Simpson argues, the "disengagement" of designated "cultural quarters" "from the city . . . underscore the fact that they are meant to be displayed as precious objects, to be admired from a distance for the local population, and that ultimately they will serve transient residents and tourists" (2017, 133). But far from only the result of corporate invasion, these strategies were written into planning law itself. In Ireland, as critical geographer Michael Byrne tells us, "the simplification of the planning system (i.e. 'fast track planning') involves a disembedding of real estate development from local contexts. But this also implies that local knowledge and local political connections become less important, raising interesting questions in terms of the local growth coalitions which are central to Irish political and economic life" (2016, 43–44). Fast-track planning continues to characterize the remnants of the creative city in Dublin, representing the inverse of political programs to put spaces back in the hands of urban communities through locally led, and publicly supported, creative and cultural projects.

A precursor to contemporary cultural regeneration schemes, Richard Florida's influential "creative class" thesis argued that cities should attract a certain kind of worker in the "creative industries."[9] The creative class, in Florida's conception, constitutes anyone from arts to finance workers (Florida 2005, 34), arguing that collaboration between these disparate industries builds more robust urban economies. This then regenerates neighborhoods and creates a trickle-down effect for workers across creative and other sectors (2005). Scott, in his analysis of "cognitive-cultural capitalism," argues that this is a centripetal force, bringing capital into city centers through intensive processes of incentivized development (2008, 6), like the Irish state's broader remit of attracting FDI to develop the wider economy. Geographers Phil Lawton, Enda Murphy, and Declan Redmond trace the resilience of the "creative-class" and "creative-city" ideas within the DCC's planning policies, whose emphasis on competition within creative enterprise "can be seen as a reinforcement and extension of neoliberal policies in Dublin since the economic downturn" (2014, 190). The policy logics, promotional materials, and spatial strategies of Irish state agencies and the industries that interact with them entangle the creative industries in a messy network of public and private interests, deliberately conflating these various activities and their ability to attract capital and generate jobs.

These extend far beyond the DCC, into the policies of state agencies with no apparent role in city planning, seen, for example, in the Fine Gael Government's Ireland 2040 and Creative Ireland policy frameworks, the latter of which featured "a five-year plan to put 'creativity at the centre of public policy'" (Slattery 2017). Creative Ireland especially placed great emphasis on the "audiovisual" industries and their global growth potential (Department of Culture, Heritage and the Gaeltacht 2017), featuring three primary policy pillars via an Arts, Film and Investment Unit: "Capital Support Schemes and funding for Arts and Culture Organisations; Film including Section 481 of the Taxes Consolidation Act 1997 (Film Tax Relief Scheme); Public Art" (Department of Culture, Heritage and the Gaeltacht 2015). It is a very broad, but very revealing remit: capital support and tax funding for the arts and media, with public art central within urban spaces of redevelopment and "cultural quarters." But these logics are also present across the board, in cultural policy agencies like SI; the corporate strategies of EI; the conversations of managers at film companies or marketing firms; or in planning documents themselves.

These logics, however faulty, are pervasive—as the story goes, "creative" workers want to be in the center of the action, in creative cultural environments, surrounded by opportunities for sharing, networking, collaborating, and consuming. This ethos drove both company policies and office designs—coworking spaces, games, free coffee and beer for socializing—as well as choice of location for such companies.[10] Major Irish-based media companies like Windmill Lane reproduced these kinds of organization and management. At Windmill Lane's postproduction offices in Dublin city center near the Docklands, I had coffee with a line producer in their open-format kitchen and dining space, with couches, a TV, and a roof deck, a space designed for chatting, meeting, sharing, and hanging out. However, the facility was largely empty, rows of computers and editing bays empty while the studio was between projects. I was told that although Screen Scene, another postproduction company, had not located nearby for any particular reason other than centrality, there was a "well-trodden" path between these offices due to the sharing of contracts, freelance workers, and clients, different laboring formations bobbing with the ebbs and flows of postproduction gigs.

But this synergy was not due to intentional "industry proximity" (Vonderau 2014) as the abiding logics of what get called "creative hubs" tend to privilege (Gill, Pratt, and Virani 2019), as much as the result might be the same. The creative city rests on the assumption that a monolithic class of "creatives" living and consuming in the city is essential to build a thriving social, cultural, and economic environment. But seeing so-called "creatives" as a monolithic unit—and a consumer-based vehicle for spatial regeneration—from the level of planning ignores class and other stratification within and across given "creative industries," which include anything from finance to filmmaking under Florida's capacious classifications of their workers (which notably exclude traditionally blue-collar labor). While there have been scholarly attempts to reclaim "creativity" within the wider productive processes of the creative industries (for example, Mayer 2011), this is nonetheless a category that was strategically instrumentalized by planners and governing bodies as not just classifying workers but as a basis of urban consumption.

However, the creative class thesis was durable in an era of expanding "convergence" (Jenkins 2006)—even media policy became increasingly written for a wide and often vague subset of "creative" enterprises. As a PricewaterhouseCoopers report from 2016 characterizes the media industries, they can include anything from "television, film, advertising, publishing, music, internet, video and online games, radio, sports, business information, amusement parks, casino gaming and more" (1).[11] The luring of this "privileged class of consumers" imagined to populate these industries, the supposedly educated workforce prepared for these precarious fields of work (see Dickinson 2024; O'Brien, Arnold, and Kerrigan 2021), was designed to accumulate reliable traffic and capital in certain ideal zones, in doing so attracting investors "who see the city not as a living space but as a landscape for profit maximisation" (Lawton, Murphy, and Redmond 2014, 194). To these investors, a marketing executive is more valuable than a film lighting technician, although both work within the media industries broadly defined within state policy and planning. The geographies of creative and media industries follow routes of capital, including those drawn by the circulations of property-driven finance, rather than draw those labor markets *to* somewhere specific. The "culture" and "creativity" cultivated by these transnational economies is always thus compromised by the fact that they can only parasitically latch onto the cultural life of a chosen place, wringing locations and their built environments dry for profit while subjecting laboring populations and "noncreative" residents to the exploitative and value extractive practices of a global productive marketplace.

A central claim, then, that I am making is that formative concepts of urban geography can be productively expanded to understanding the intersecting affective, cultural, and labor dimensions of life in the "creative city," a spatial policy which was central to the "experimental" landscape of post–financial crisis urbanism, as Cian O'Callaghan and Cesare Di Feliciantonio argue (2023). These experimental processes must be understood as both experiential and systemic. Scott has usefully theorized "cognitive-cultural capitalism" as the driver of contemporary urban development, tracing a history of a widely and variously theorized "new economy" of so-called post-Fordist modes of production in the late twentieth century, which have centralized in cities and are driven through intertwined economic and cultural activity (2008, 11–14). But it is useful to speculate on the dimensions of creative labor encapsulated within post-Fordist production and its organization and management across supply chains through the affective environment of financial crisis and its aftermath. In an environment of widespread job loss and austerity, as has characterized post–financial crisis Ireland, everyday life changes drastically, as the structural precarity of financial and logistical systems was laid bare by the recession of capital. The financial crisis, rather than untethering Ireland from a world system to which it had become an (unsympathetic) victim, came to signify an intensification of these variously theorized modes of production and their spatial distribution. FDI, in an apparent paradox, was posed as a necessary ladder out of the depths of this neoliberal crisis (whether by internal or external logics) and thus plugged the country into even more concentrated media and technology supply chains.

The redirection and redistribution of material flows, of capital's migration from one sector to another in times of crisis, are operationalized effectively through the built environment. This occurs through planning and zoning designed to reshape the space, which, intentionally or not, leads to the displacement and transformation of communities, cultural practices, and industries. Ireland and NAMA in particular were tasked to deal with the empty investment of what David Harvey has called the "spatial fix" after the financial crisis. The spatial fix describes the tendency during crises of overaccumulation to expand investment in property markets. By Harvey's definition, the spatial fix deals with excess capital in a particular way: "Spatial displacement entails the absorption of excess capital and labour in geographical expansion. This 'spatial fix' . . . to the overaccumulation problem entails the production of new spaces within which capitalist production can proceed (through infrastructural investments, for example), the growth of trade and direct investments, and the exploration of new possibilities for the exploitation of labour power. Here, too, the credit system and fictitious capital formation, backed by state fiscal, monetary, and, where necessary, military power, become vital mediating influences" (1990, 183). In a post-crisis environment characterized by a *lack* of new investment and an *excess* of sunk capital into built spaces, the state sought out new flows of capital to financially reanimate the built environment and reactivate the growth that had stalled—including by introducing special incentives like a 32 percent-plus tax break for media production, SDZs, enthusiastic state partnerships through NAMA, and long-standing industrial policies such as the 12.5 percent corporation tax rate. These were innovated long before the crisis, but were further entrenched by perceived necessity. While offering short-term growth models, they also deepened dependence on fickle flows of capital, which could recede at any moment, corroding the cultural life of a place via enforced precarity, all in the name of "creative" investment in the built environment and labor market.

This "spatial fix," or "creative fix" in this case, demonstrates the degree to which the hold of the "creative class" logic became stronger in Dublin after the downturn (Lawton 2013, 108) and its privilege in city planning disproportionate. Capital interests (facilitated by the DCC) determined growth, not, as creative city planners would lead us to believe, the self-directed internal migration of labor into certain sectors. Investments in space in this creative sector were a different kind of spatial fix *operationalized* through so-called creative labor and enterprise, succeeding in encouraging global competitiveness across disconnected spaces and forming new spatial markets for surplus capital (Lawton, Murphy, and Redmond 2014, 194). This experimental landscape enabled by austerity also risked crystallizing these structures, wherein the exceptional deregulatory measures become locked into infrastructures that support naturalization of permanent austerity. Lawton warns that culture-led regeneration tactics, "while successful in filling a void and, to a certain extent, challenging official meanings of place, the temporariness of such measures highlights the degree to which they are at the mercy of property market fluctuations. There is a very real danger that without the proper structures in place, the positive role of such initiatives will be sidelined at the first sight of a return to a 'normal' property market" (2013, 109). Community-led cultural groups thus were frequently and rightfully averse to partnering with these creative initiatives, even as their activities conditioned spaces for status as "creative" districts with vibrant cultural life. For example, the Creative Spaces Collective lost its footing in the city center of Dublin, as Smithfield, a largely artist-led regeneration area, became one of the more gentrified in the city in the mid-2010s (see Guinan 2016). "Creatives" act as the foot soldiers of spatial change, only to be displaced themselves once space resumes profitability by redirecting flows of capital into the built environment and certain strategic industrial sectors.

These fixes are thus only temporary solutions to the endemic crises of financialization and can only lead to new ruin once capital recedes. But to speak of a creative fix, we have to recognize that this inescapably *does* instrumentalize culture and the affective responses to financial crisis in ways that Marxist autonomia and post-autonomia thought has long tied to the growth of "immaterial," "affective," and "free" labor in contemporary creative industries, and the various ideologies that underpin these logics of "flexible" work (Federici 2013; Gill and Pratt 2008; Lazzarato 1996; Lovink and Rossiter 2007; Terranova 2000). This specific crisis of—and response to—overaccumulation thus has something more explicitly to the spatial instantiation of creative industry politics and its normalization of precarity, which is experienced through the constant changes of the built environment in the creative city (Lawton, Murphy, and Redmond 2014, 194). The "fix" for this crisis involved the reorientation of entire corporate strategies of capture and value production, enacted in and through space. This included the harnessing of social cooperation "between brains" through so-called "cognitive capitalism," in person, across territories, through networks, and along supply chains (Marazzi 2011, 57). One of the primary spatial avenues for this was through media and technology industry production.

The Visual Media of Planning and Development

As suggested in chapter 1, visual culture is central to how planning and development operate, whether images, promotional materials, or plans and maps, and whether directed toward outside investors, the public, or locals. These materials represent both the ideology and practice of the creative city and its forms of value driven by property finance and just-in-time management. As geographers Gillian Rose, Monica Degen, and Clare Melhuish have emphasized, promotional materials are important for how these spaces operate and create themselves, whether by advertising tax incentives or certain lifestyles to investors and consumers (2014). These promotional methods—and the strategic benefits they advertise—project and reorganize territory in particular ways, but are also industrial formations in themselves, with architecture and design firms "creating" digital environments or "atmospheres" to be enacted in the city. Degen, Melhuish, and Rose argue that the "importance of atmosphere as an economic value informing the production and consumption of consumer goods means that the boundaries between artistic work and industrial work are progressively blurring" (2017, 7–8). The designers and marketers of these materials, much like city planners, perform the logistical work of preparing spaces for circulation, imagining a particular future for capital accumulation. The plans, visual culture, and real environments of the "creative city" paint over the everyday violence of financial and logistical capitalism in urban spaces, which are laden with the extractive and disruptive mechanisms of global capital, especially in a space as driven by FDI as contemporary Dublin. This section argues that these *regimes of visuality* drive and organize imaginations of circulation, spatial development, and politics, representing these future-driven logics of spatial use and governance through visual culture.

Buildings and space, however producing value within finance and creative industries, are also visual referents of these overlapping economies. As in the case of ghost estates, this visual component contains powerful cultural significance for different political experiences and agendas. As anthropologist Llerena Guiu Searle argues, speaking from the context of neoliberal real estate development in India, "new buildings are predicated on *forecasted* social and economic changes. Concrete and steel obscure the stories about India's growth that fuel construction. Once completed, buildings become evidence that these stories were true" (2016, 2). Development projects are enacted top-down, but must navigate spatial and community politics and histories to move forward, which stand as barriers or boosters to their materialization. The part-time film studio-cum-apartment block plan for a former Player Wills Factory in the Coombe, a NAMA property controversially earmarked for redevelopment (Thomas 2019) and frequently used by film and TV productions in the 2000s and 2010s, was partially objected to because of its proposed height potentially disrupting the area's "visual amenities" (see An Bord Pleanála 2021). The "visual" itself becomes a resource for certain kinds of property value logics.

The visual culture and designs of urban spaces narrate a particular "story" about the city, effectively "greasing the wheels" for circulation through it (Dickinson 2024). In doing so, planning images perform *logistical* work in preparing spaces and subjects for capitalist spatial futures. In this section, I use two examples of development projects in the Docklands to emphasize the designs of redevelopment at play through promotional materials and "gray media" circulating about them (Opaque Media 2017). These media forms imagine a smooth and stable future city. However, the *tumult* of the creative city's constant construction is more characteristic of its experience. Just as logistics manages the "turbulent circulation" of the world economy (Chua et al. 2018), these spaces of the creative city, however much they are designed to direct consumer and business traffic into ideal zones of work and leisure, are chaotic, sites of constant construction, and evidence of the violent rationalities of displacement that drive them. The Docklands are not only *former* industrial lands—they neighbor existing working-class communities and active industrial and port and logistics operations. An everyday observer can visibly see the ways in which these developments continue to infringe upon existing communities and structures, as spaces of former employment and ways of life are subsumed under the imagery and space of the creative city, displaced by corporate "creativity" that supposedly takes care of the public culture of place.

On a tour of the Docklands with scholar and artist Paul O'Neill in winter 2019 (for an overview, see Humphries and Kapila 2019),[12] he had invited an artist, student, and Ringsend resident (a working-class community neighboring the Docklands), who described the ongoing attempts of multinationals to expand their operations into the neighborhood. Even without these attempted buyouts, however, he admitted that he would never himself be able to afford to live there, due to soaring property prices in Dublin's widespread housing crisis caused in part by the predatory presence of these companies, cynically nicknamed the "Google effect" by the national press, especially in working-class neighborhoods like Ringsend and Irishtown (see Gallagher 2017; also O'Neill 2019). Creative city development projects like those in the Docklands, like FDI-driven development projects more broadly (see chapter 4), justify their private partnerships by "ripple" and "spillover" effects in surrounding communities and industries (see, for example, Tom Fleming Creative Consultancy 2015; also chapter 3). However, while there is undeniable "spillover" happening in terms of the spread of spatial regeneration, any promoted "benefits" to local economies (Thomas 2017) actively occlude the financial and property practices that make Dublin unaffordable for working- and even middle-class residents—just another part of the corporate sleight of hand that precedes outright displacement.

This Ringsend resident focused especially on a development called "Capital Dock," a mixed-use tower project in the Docklands funded by Los Angeles investment firm Kennedy Wilson, a joint venture with NAMA and Toronto company Fairfax Financial Holdings (Hamilton 2017). The young man told us that as a youth he used to wander this formerly vacant lot with friends during the financial crisis, but had recently been accosted by security guards when visiting for a nostalgia trip. This space, he said, was no longer for longtime local residents, even though it had been essentially abandoned by the state and capital for years beforehand. The journey from ruin to financial fortress was a rapid one. At one point prior to the crisis, during the "Boom," the site had been earmarked for a tower project development backed by Irish band U2, which would have become the tallest building in Dublin, complete with U2 practice spaces on the top two floors. When that proposal fell apart following the financial crisis, the land was vacant for years until Capital Dock piggybacked onto the existing site. In this way, rather than activating or accounting for existing community relations with the site, the developers (including NAMA) simply reanimated a stagnant plan for the property, demonstrating the favorability of global capital within such procedures.

When looking at sites such as this, there is the visible instantiation of the building and construction site on the city along with the histories, visual culture, and lived experiences coursing through the site that are available for immediate study. Out of frame, however, there are also the materials being shown to investors, buyers, and boardrooms, which provide insight into the systemic functions that constitute these spaces. Each of these visual and experiential referents reveal the deep logics at work, and give a better sense of the variety of actors and processes involved (see figures 2.5 and 2.6). A promotional video for the site (Kennedy Wilson 2016) exemplifies the "gray media" of corporate promotion circulating in a particular wealthy ecology of investors, planners, and high-class consumers (Opaque Media 2017).[13] Beginning with juxtaposed images of sped-up construction with leisure activities like water sports in the Grand Canal, the video presents "creative" geographic and demographic imaginations of the future living and working environment, as it would be a "destination" for leisure, a "truly networked urban quarter," "a vibrant hub"—corporate buzzwords all associated with the creative city. To support foreign investors with a local voice, the video is narrated by one Declan Kelly, in a hard hat, explaining the virtues of Kennedy Wilson. Kelly praises the project's speed, attributing it to

2.5 Capital Dock rendering in a promotional video (Kennedy Wilson 2016, 1:10).

2.6 Declan Kelly, regional director for a construction firm working on the site, in front of the ongoing project (Kennedy Wilson 2016, 1:36).

Kennedy Wilson's capital-securing capabilities—rather than the Docklands SDZ, which fast-tracked planning and construction.

Visiting the area frequently, I noticed graffiti and commissioned murals—usually on disused industrial spaces—blending and existing side by side with digital renderings of future environments such as Capital Dock as seen in the corporate video (figures 2.7, 2.8, and 2.9). These design and artistic visions are frequently clustered in "cultural quarters" amid their regeneration logics, which differentially distribute the public and legal legitimation of certain kinds of art over others in terms of spatial use. For example, street art collective Subset created a variety of mural cycles in the late 2010s, such as the "Grey Area Project," in protest of prohibitive and exclusionary public arts policies (Mullally 2018). As Una Mullally emphasizes the contrast between publicly oriented planning and the protests around corporate development, "Street art moves quickly, local government moves slowly. Creative endeavours are inherently dynamic, local government often seems inert" (2017). But there is also a strange overlap here between the ad hoc visualities of street art and the rapid pace of building and regeneration by developers once they are handed the keys to the city—one operates underneath legislation and legitimation, and the other above, coexisting in ostensible opposition but going up in tandem. By animating what they called "gray areas" of the city and its use, Subset protested both the prohibitive planning policies and reimagined different kinds of urban use. However, what Mullally's analysis glosses over are the competing but compatible visualities at play, perhaps serendipitously or ironically mirrored across the "gray media" of corporate planning and the "gray areas" of actually existing urban space, everyday legalities, and public use. These ad hoc visualities of the "creative city" are as much a product of its contradictory presence as the speculative gazes of multinational capital.

Finance, and its modes of spatial operation and security, can be visualized through such environments but can only be understood through the histories and processes it masks. We can see how these logics are visualized and enacted through the experience of the creative city, at both intentional and accidental levels: whether in the plans plastered on construction hoardings, or in the contingent forms of proximity and consumption that characterize everyday interactions within such spaces. Each is designed to enclose, manage, and capture value through circulation, conjuring new conduits for flows of capital. The chaotic jumble of construction, walls, environments in various states of assembly and repair, are fascinatingly contrasted with corporate media: whether digital renderings of these future

2.7 Construction hoarding around Capital Dock in summer 2018. Photo by Patrick Brodie.

environments on walls and hoardings, or in the policy and planning documents circulating in boardrooms in high-rises, offices in Sandyford, and state buildings. These images project shiny buildings, young families and happy consumers, businesspeople and workers coexisting within controlled and idealized environments. Creative subjects are reduced to consuming subjects, visualized by normative marketing assumptions that betray who the space is really for.

The interaction between these controlled representations and the actually existing contingency of the built environment are crucial to how the story and experience of the "creative city" were told, and should be taken seriously, even if their future visions were (and remain) disagreeable. Searle issues a polemic that "scholars on the political Left have trouble taking capitalists' worldviews seriously for fear of appearing to condone them. . . . Moreover, [David] Harvey and Marx suggest capitalists' understandings are distorted by the structures of capitalism itself: bankers cannot 'see' what's really going on behind the surface appearance of money begetting money" (2016, 12). But, she continues, "capitalists' stories are not descriptions only to be judged right or wrong; they are 'efforts to re-make the social world.' . . . When people tell stories, they don't merely exchange information; they create social relationships with one another and conceptually organize the world so as to act in it. Thus, instead of debating whether financiers are delusional or prescient, it is more productive to understand what their stories *do*" (12). Rather than ascribing value judgments to such plans themselves and their visual renderings, it is perhaps more productive to analyze these stories *as stories* in order to disentangle the political logics that tell them and then enact them into being. By engaging with how capitalists visually narrate the creative city, we can actually pinpoint not only the tools by which these stories are told, but the points of friction and blockage that these tools are always in conflict with—whether graffiti collectives or, in the case of Airbnb's former headquarters behind Capital Dock, crisis-era arts spaces like Mabo's (Posor 2014).[14] These competing materialities and visualities frequently showed parallel stories that interacted with financial and corporate development in symbiotic or conflicting ways—and these points of meeting, especially between cultural activity and enterprise in the austerity-driven "creative city," remain deeply instructive.

Geographers Eoin O'Mahoney and Phil Lawton analyze the brand of "entrepreneurialism" at the heart of such urban visualities (2019) to understand the pervasiveness of visual architectural renderings and their marketing of urban space. Crucial to how they work is a future-leaning ethos, imagining a future free from ruin, as media scholar Joshua Neves argues: "The alchemy or projection at issue here is both the transformation of everyday space into an image and the temporal delay animated by urban renderings: projections thrust and defer the present into the future. In short, design's visual culture permeates everyday life by consolidating an aesthetics of development or progress . . . and reinforcing neoliberal modes as the inevitable pathway to development—a term that itself conflates modernisation with the improvement of standards of living" (2013, 293). But particular to Dublin are the ways in which a particular past is thrust into the future, bypassing the ugly present (of displacement, construction, and surrounding ruins). They also construct a particular, and in some ways total (or, if not, deeply pervasive), social vision of the city. As O'Mahoney and Lawton argue, "While the developers act predominantly in a self-interested manner so as to differentiate their development, a form of collective image-making can also be identified in distinctive urban areas. Furthermore, although largely adhering to what can be seen as a globalized form of image-making, the hoardings are also highly localized, drawing both upon distinctive mythologies of particular parts of the city and also seeking to reproduce these mythologies, albeit repackaged through the selling of the city as both sensitive to the past, while also looking to an imagined future" (2019, 203). This future-leaning, backward look—with minimal recourse to the ruinous present—was indicative of Irish economic development post-crisis more broadly, whether in media or cultural tourism (discussed more in chapter 3), simultaneously erasing inconvenient histories, terminal presents, and alternative futures. But specifically, these visualities—glossy digital renderings within the most financialized and consumer-driven zones of urban areas—coexisted with history and the contingencies of local culture and even dissent, whether they intended to or not.

Crucially, and ambiguously, this was not necessarily a negative for the ideals of creativity and culture. Graffiti, posters, murals, and artwork, official or not, on ruined and overgrown, under-construction, or new spaces were incorporated into the visual logic of the creative district (figures 2.8 and 2.9). These competing but compatible visualities—contingent or dissenting, however manufactured—worked together to enclose possibilities for real resistance against regeneration. The lived experience of these spaces—as hard to study as they were, due to their ad hoc and transient character— were captured within the imagined encounters that they created, and the ones in the past (and the alternative futures) that they foreclosed. Spaces

2.8 Tag on Boland's Quay design hoarding in winter 2019. Photo by Patrick Brodie.

2.9 Public art promoting Benson Street Park in summer 2018. Photo by Patrick Brodie.

plastered with these images were as melancholy and pregnant with evacuated meaning as ghost estates—futures that were on the cusp of emergence,
foreclosed by the ruinous circulations of capital.

The Boland's Quay mixed-use development project in the Docklands
also contained an array of histories and newly developing "stories" told
by developers. These were not actively occluding existing and past spaces
and social forms—they also demonstrated the mechanisms by which culture and heritage were (re)activated within spatial development via affective promise. Boland's Quay was another "flagship project" for NAMA in
partnership with global real estate giant Savills and Bruce Kennedy Doyle
Architects, among others. It promised a "redevelopment" of the historical
landmark Boland's Mill, known both for its industrial and revolutionary
heritage as a significant portside mill and a major combat site during the
1916 Easter Rising. As reported in the *Irish Times*, "The historic buildings
on the site, dating from the 1830s, will be 'sympathetically restored' and
their heritage preserved adding to the 'architectural uniqueness of the
development,' the NAMA-appointed receiver Mark Reynolds of Savills
said" (Gleeson 2018). In a video promoting the project (Image Now n.d.),
NAMA executive Frank Daly appears as a talking head: "Because we're a
state body, we don't make commitments lightly." The biggest sources of FDI,
he says, "[are] here, but they need more space." Over a dramatic montage
of aerial and everyday shots of the Docklands, the video advertises the
project's proximity, "right at the water's edge," to the various "household
name" multinational tech, media, and finance companies located in the
Docklands (figure 2.10), "reflecting the past, embracing the future in the
most vibrant hub of the city." In a manner familiar from the Capital Dock
video, developers, architects, and demolition experts speak about the site's
uniqueness, industrial heritage, and protected structures, which will be
"renovated and brought back to new environment [*sic*]," brought "back to
life" by "making new interventions" with design and property investment.
Toward the end of the video, an aerial map of the Docklands shows new
structures plopping down on the quay, colonizing the space with new edi-
fices falling from the sky, indistinguishable in style from existing buildings
(figure 2.11). This bird's-eye view, characteristic of colonial mapping and
militarized imperial sight (see Parks and Kaplan 2017), coincides with no
reference to the anti-colonial revolutionary history of the site, but it ends
with an unintentionally pertinent statement: "Available for occupation
2018." Google's purchase of the site for €300 million in 2018 puts a bow
on the neocolonial irony (Gleeson 2018).[15]

2.10 Boland's Quay, with all surrounding major FDI firms (Image Now n.d.).

2.11 The mold of Boland's Quay parachuted onto existing territory (Image Now n.d.).

The producer of the promotional video, Image Now, is a Dublin-based brand consultancy that manages creative projects ranging from ad design to stage design to video production, ranging across a variety of creative endeavors and countries (primarily the United States and Ireland). What is perhaps most striking in the video is that the protection of structures and spaces of heritage—through state organizations like DCC and schemes like SDZs—is directly handed over to these industries under the guises of culture and regeneration. These images thus brand and articulate themselves within a "global economy of images" (Degen, Melhuish, and Rose 2017), and their specific urban aesthetics are intricately entangled with their places of expression.[16] A page in the brochure for the development advertises Boland's Quay as a place "where business and culture meet" (Boland's Quay n.d., 19). The hoardings around the site were blazoned with the slogans "Reflecting the past. Building the present. Embracing the future" (figure 2.12). In this story, the past is a refracted vision of the present- and future-driven enterprise of regeneration. With Google having since bought the space, it is much more obvious what kind of future was *actually* being built.[17] Their prior office spaces can be seen in the digital design images featured in the promotional video (figure 2.13), overlooking the company's next property asset.

The media professionals at Image Now were hired by semi-state and corporate organizations to tell this project's chosen story to potential investors, and to a public perhaps skeptical of its location on an existing heritage site. This is not to suggest that such enterprises (or their workers) are solely responsible for the visual culture of such developments—rather, they are a cog in a wider supply chain conditioning spaces for future circulation. The presumably limited-term nature of this contract, through which the brand consultancy and its workers produced visual culture and promotional materials—from this video, to an investors' brochure, to press releases—demonstrates the degree to which supply-chain management across a variety of scales and businesses is what makes spaces operational for capital, whether we are speaking of media industries or construction enterprise.[18] As one cog in the machine of the creative city, Image Now performs a vital service acclimatizing actors peripheral to how we imagine supply chains and development to function. Seeing Image Now as a media company, based just a stone's throw away from the Docklands (and near Windmill Lane's offices), shows how the expanded remits of the media industries suggested by this book have transformative roles within spatial development beyond conceptions of proximity and worker mobility. Media

2.12 Digital rendering of the project next to the iconic façade in winter 2019. Photo by Patrick Brodie.

2.13 Rendering from Image Now video, with Google's prior offices at middle right (n.d.) Photo by Patrick Brodie.

companies often operate within the very same supply chains that create the financialized creative city.

Whatever the political economy of such media enterprises, they are tasked with producing a visual culture that tells the stories with capital as the protagonist, endorsed by the state, and harnessing or excluding existing communities and histories. In particular, in the case of Boland's Quay, we can see the role of visual culture in packaging heritage and design within the logics of finance (particularly property) capital. Such images act as both product and propaganda of planning and development, complete or incomplete. In particular, the visuality put forth by digital renderings of these environments reclaims these spaces *for* capital by claiming actualization within spaces devoid of existing relations. Neves contends that these images "abstract, make static, defer, and rely on the bird's eye view" (2013, 296), even though each navigates through a political economy of actors on the ground to enact such a worldview. There is a variety of perspectives offered by these projections, at least in the physical sense. We may say that eye-level visualizations do nothing but transmute the violence of the bird's-eye view to a more familiar, everyday space, plastered on the barricades around construction sites. However, this move should be analyzed in terms of the strategic points of view put forward by design firms and planning bodies and the problematic lineages of their aesthetics and then exploded across the chaos of urban space under construction. Because as Neves also argues, "city plans and urban landscapes are decidedly void of human life" (2013, 296). Echoing the erasure of life and labor from early FDI imaginations of the Irish landscape promoted by the state (see chapter 1), the logistical imagination of a "smooth world" (Cowen 2014), colonial geographies of terra nullius (Yusoff 2018), as well as those of data centers' natural vistas (see chapter 4), this visual culture expresses the pervasive politics of visibility/invisibility at the heart of finance, logistics, and infrastructural development.

But as Cowen has made clear (2014), management's "logistical revolution," which this book argues has come to influence much wider economic and governing systems, means that commerce and security, property and territory, visibility and legibility are increasingly inextricable in practice. Whether in the geographies of creative work that function within much wider supply chains or the violent logics of displacement promoted by spatial visualization, the turbulent logics of circulatory capitalism can be read through the stories that capitalists told about the creative city. These logics of futures, reuse, and redevelopment, where the future city was already on present display, where former factories became impromptu or permanent film studios, port lands became mazes of glossy and graffitied hoardings, and heritage buildings became big tech offices, were seductive and difficult to short-circuit, despite how apparent the exploitative and extractive practices of the state and FDI might have appeared within even a cursory study. It was impossible to tell what would go forward, what would transform, what would be foreclosed, and what would be disrupted through these analyses.

As you read this book, many of these futures remain in progress or permanently canceled. But this is why it remains so important to locate points where negotiations about spatial and cultural futures are still ongoing. In the remainder of this chapter, I outline various projects that were involved on the Poolbeg Peninsula, focusing especially on the Poolbeg West SDZ and the Pigeon House Power Station and Hotel, both of which found themselves in the sights of planners and investors as places for film production and creative enterprise.

Reuse and Alternative Media Practices on Poolbeg

While the Docklands have been filled with cranes and media since the early days of the "recovery," a new regeneration frontier just east of the SDZ only began to be filled with these speculative media in the waning years of the 2010s—the Poolbeg Peninsula, with the working-class neighborhoods of Irishtown and Ringsend sandwiched in between. The Poolbeg landmass, an industrialized peninsula highlighted by the iconic Pigeon House Power Station smokestacks,[19] a famous lighthouse at the far eastern tip, and the scenic Sandymount Strand buttressing its south end, was a significant site of imagination for Dublin's spatial future during the recovery.

At the end of 2019, the Poolbeg West SDZ was set for development as a mixed-use complex by multinational Deloitte. The plans were widely contested. The site initially entered the news in 2016 because of the former Irish Glass Bottle and adjacent Fabrizia site, in the form of a land grab featuring the aforementioned proposal for Dublin Bay Studios. Lobbied by Bono and supported by the government and DCC, investors were nonetheless competing with a more familiar mixed-use property proposal, which promoted waterfront living and commercial activity in view of the famous Poolbeg skyline. The site was ultimately picked up for housing development to "fast-track planning for up to 3,000 homes on former industrial lands" (Kelly 2016). At this time, DCC owned a portion of the site due to its absorption of the share owned by the now-defunct DDDA, and the scheme was also being administered by NAMA as a partial, developer-led attempt to remedy the prevailing housing crisis in Dublin, with a stated focus (nearly 30 percent) on social and affordable housing. In familiar fashion, demonstrating the class demographics favored by such development's imaginations, upon acquisition of the site Deloitte disputed the social and affordable housing requirements, reneging on an official agreement to provide 25 percent social housing, citing the citywide standard of 10 percent (Kelly 2018).

Developing on and around the port is complicated, and there are competing interests: From global media production to local demands for housing, plans are subject to the development interests of organizations like DCC and the DPC, the latter of which was so blunt as to call the film studio plans "daft."[20] The DPC at the time was balancing a newly postindustrial role as property asset manager, arm in arm with NAMA but also in line with resource agencies like Coillte and Bord na Móna (see chapter 4), with ongoing operations as the country's largest port. The semi-state company expanded its remit as a cultural and planning agency, aligning with the "creative" remit of the encroaching Docklands.[21] A draft Poolbeg West SDZ planning document emphasizes these multiple roles: "The challenge is to provide commercial and residential uses with a good level of amenity, while at the same time allowing Dublin Port to continue to operate and perform its vital trade route function for the State" (Dublin City Council 2017b, 19).

Poolbeg represented a unique challenge for "Docklands" expansion, as a polluted, detached industrial hinterland that was just barely close enough to strategic and well-connected areas to merit special interest. Despite separation from city infrastructure and an unappealing environment—during my research, public transit was scarce, and the area was mostly loud, smelly, and rubble-strewn—the peninsula's location as the last frontier for the Docklands, industrially and physically, was frequently referenced in city planning materials:

> The key growth sectors also relevant to a sustainable city quarter are Information and Communication Technology (ICT), Financial Services, Consumer & Business Services and *Content Industry*. These sectors have a significant presence in Docklands, save for the latter where there is a small but growing presence of the industry subsectors such as *media, music, film, digital and games industry*. It is considered that Poolbeg West can assist in ensuring that the Docklands and the city continue to attract high value economic activity in key growth sec-

tors across a range of corporate entities from start-up to large mature organisations, indigenous and multinational enterprises. A core part of this challenge is ensuring that a range of employment options and skills needs become available, creating opportunities for the local community. Dublin City Council, through the Docklands Forum, intend to work closely with education, training and other agencies to ensure maximum opportunity and benefit from employment opportunities for residents of the wider area. (Dublin City Council 2017b, 20, emphasis mine)

After describing the logic through which media industries must be clustered with these other industries, the DCC extended this idea into the support for the film studio: "Dublin City Council supports uses associated with media/digital media and film production. Dublin City Council will actively engage with all the relevant stakeholders to explore the opportunity for a Film Studio and associated uses to locate within the SDZ commensurate with the need to first and foremost maximise housing delivery and secure the strategic development of Dublin Port as Ireland's most strategic and largest port" (21). The connections between the state's media policy and the private sector—in the form of this never-to-be-built film studio—was written into the city's future plans.[22]

That said, this idea was significantly and interestingly amended in revised planning documents (Dublin City Council 2017a), which doubled down on social and affordable housing requirements after appeals from Deloitte, reinforcing that 900 of the 3,500 housing units to be built must be social and/or affordable. This was a basic provision to ensure that people from Irishtown and Ringsend might still afford to live in their own neighborhood. However, the plan still left open a loophole frequently used by developers in public/private agreements: The prevailing method of providing social and affordable housing was to sell the DCC these units offsite, meaning that communities that were meant to be cared for within these plans—locating them in or near their existing neighborhoods—would be forced to relocate far away or even outside of the city (Reddan 2019). This highly systematized, violent tool of gentrification publicly zoned land to sell its development rights to investment funds and private developers—expropriation at work, promoted as economic progress in the creative city.

Resident groups were (and remain) active around these issues, with additional productive conversations to be drawn between groups with apparently diverging aims. For example, the Creative Spaces Collective advocated heavily for more cultural spaces in Dublin outside of typical redevelopment logics and were part of a campaign to secure forty artists' studios in the plans for the Poolbeg West SDZ. The Sandymount and Merrion Residents' Association lodged complaints, arguing that the area needed improved infrastructure and environmental protections before any development could happen (MacArthur 2017). More radically, the Irish Glass Bottle Housing Action Group mobilized against "NAMAtown" (Irish Glass Bottle Housing Action Group 2019) and negotiated for the 900 guaranteed social and affordable housing units through the scheme out of a total 3,500. While these groups' aims may appear at odds, each attempted to imagine different and more locally beneficial spatial use from that of FDI-driven development. In the revision to the plans, after public consultation and controversy around housing provisions and the film studio proposal, a mayor's executive report detailed DPC's opposition to Dublin Bay Studios' desire to use land they had earmarked for port activity, as well as the local community and GAA club's objection to potential displacements (Dublin City Council 2017a, Material Alteration no. 4). Other actors lodging varied complaints and proposed alterations included local community groups, political parties, film producers, film policy officials, industry boosters, arts groups, and transport unions, among others. What this amendment does is provide a basic clarification as to the integration of the enterprise role of the film studios, and the proposed built environment in general, within an actually existing local ecosystem, full of conflicting or coalescing interests.

Relatedly, the Pigeon House Power Station and Hotel was bid upon as a site for a "creative industries" cluster deep into these industrial lands, at the far edges of what could reasonably be considered the Docklands. This site has acted as an impromptu film set on several occasions, including for the 2002 Matthew McConaughey and Christian Bale film *Reign of Fire* and the RTÉ direct provision drama *Taken Down* (2018). The hotel hosted temporary offices of Irish Water and a small film production company at the time of my own visit in winter 2019. The power station, in operation through the 1900s until the 1970s, features iconic smokestacks, a staple of the Dublin skyline.[23] The hotel, a Georgian structure built in the late 1700s, is similarly historic, a place for passengers arriving at the then-colonial port. While portside, the area has been progressively cut off by other industrial developments, including the active port, waste-to-energy plant, and wastewater treatment site. Nonetheless, five companies in 2018 reportedly speculated on the site as a future hub for redevelopment (see Kelly 2019).

However, while under the speculative gaze of city planners and private investors, an artist, filmmaker, and postproduction specialist spoke with me in 2019 about his far different relationship to this space through the crisis. He had put on "run and gun" public screenings in the 2010s, where they literally broke into the space to put on one-time expanded cinema happenings as part of a series called Open Nights Cinema. These were short-lived occupations of the space for exhibition and performance. He had also carried out some actual filming there, joking that the dilapidated power station was a safety hazard—while filming with another man and his dog years before, the dog had fallen into a small pit and nearly drowned in the muck before they finally managed to rig a way to get the dog out. While he was filming, through word of mouth, he learned of a former film production company that had used the site, and even found some old used film in the mud on the floor of the building, from a lost RTÉ- and IFB-funded short program called *The Poorhouse* (Frank Stapleton, 1996) about a women's workhouse during the Famine.[24] He took this largely ruined and decaying footage, rotted away by exposure to the polluted elements, and re-edited and screened it as a performance piece, which he called *The Poorhouse Revisited* (Higgins 2012). The ghosts of the past continue to haunt the present, with different material effects and experiences.

The convergence of reuse and recycling in the Poolbeg artist and his collaborators' art practice was directly antagonistic to that dominating the practice of the DCC and other spatial planning bodies in Ireland, especially in relation to decay, ruin, and (re)use in these haunted environments. The state, via semi-state organisms and public/private partnerships, would sacrifice public heritage at the altar of profit, while at the same time commodifying culture and creativity as central motivations. The logic of organizations like NAMA and its private partners enclosed these ruins of not just capital but culture in the built environment, releasing not the potential of the outmoded, as suggested by Walter Benjamin and developed by Erika Balsom in her research on ruin art practice (2009), but violence against the obsolete and the unprofitable. Balsom instructively articulates the concept of the ruin through the politics of gallery practices on celluloid film, resonating with the Poolbeg artist's use of ruined film stock as a proxy for and within existing ruin. Balsom argues, through theorist Andreas Huyssen, that contemporary obsession with ruins "hides a nostalgia for an earlier age that had not yet lost its power to imagine other futures" (Huyssen, quoted in Balsom 2009, 421). Culture, however, as Balsom demonstrates, is not spared neoliberal logics of (re)use even within its attempted disruption or activation as a public act, but only if it remains within a certain system of value determined by a circulatory market—whether that of the gallery or of the creative city. But what if, as Balsom suggests, "the affective complex" of ruin and its "uncanny linkage of past and future" creates a different kind of atmosphere, "a [Benjaminian] aura linked to contingency and historicity" (426)—an aesthetic and even community use value rather than exchange value?

The practice of reuse seen by those who creatively repurpose vacant spaces on a one-off and nonprogrammatic basis—and ruined film materials of *publicly funded media*, in the case of *The Poorhouse Revisited*—may possibly model different and more sustainable ways of living in and moving about space outside of the development logics motivating approaches to space and culture throughout Dublin and Ireland more broadly. This is especially true of spaces in the crosshairs of corporate-led development's regimes of sight. Fugitive and unruly uses of such spaces harken to avant-garde practices of detournement, the visual and lived reappropriation of space and material for truly creative practice (see figures 2.14 and 2.15). Of course, seeing these images recalls traditional notions of the avant-garde as ideally blurring the boundaries between art and life, activating politics through artistic practice (see Connolly 2004). We can remember Benjamin's famous assertion about the politicization of aesthetics in "The Work of Art in the Age of Its Technological Reproducibility," where communism counters fascism with political art practice (2008, 42), with the tactile, avant-garde activation of a collective political subject.

But when looking at spatial development in Ireland, we can also see the feared inverse that occupies Benjamin's mind, through his analysis of Italian Futurists' fetishizing of imperial warfare: "The logical outcome of fascism is an aestheticizing of political life" (2008, 41). Through the visual culture of corporate development and state/corporate place-branding strategies, we can thus also glimpse the dangerous ways in which these public and private actors aestheticize austerity politics: the naturalization of the tumult and ruin of finance capital, shielded underneath future-driven projections promising transformation, prosperity, and the capitalist mundane. The state and its corporate partners are both equally guilty of these aesthetic politics, experimented in the ruins of the post–financial crisis city. What we can see in these competing practices, then, the cultural occupation and the systemic enclosure of space, are two truly competing regimes of visuality—that of new cultural imaginations, repair, repurposing, and

2.14 Detail from *The Poorhouse Revisited* (Higgins 2012), demonstrating the aesthetics of ruination and decay at the heart of the project—and crucially, referencing the public funding for the original film.

2.15 Photo from a showing of *The Poorhouse Revisited* (Higgins 2012).

fugitive occupation; and that of capitalist green- and brownfields, experimentation, innovation, regeneration, and revitalization. Each of these
regimes has its own demographic, cultural, and political imagination of
what a just spatial future would look like. While we can obviously critique
how the former imagines the artist as the harbinger of social change and
remember the impotence (and often enthusiastic participation) of art in
the face of massive systemic apparatuses (Steyerl 2017), this is nonetheless a mode of contestation, modeling reclamation of space for people and
culture from capital and profit. At the very least, it is expression of and
through the cultural life of a place, liberated from exchange-value, unen-
closable within an economy of forms that values property and profit over
culture and livelihood.

Conclusion: Culture, Capital, and Struggle

Planning in Dublin contains ghosts of the crisis even as its neoliberal policies march onward still. However, these ghosts do not exist only in the
built space, but also take shape in the subjectivities of those directly affected by austerity and financialization in the guilt of collective debt driven
by supranational regulators. One must not forget that the crisis was one
of massive debt *in particular*, a structure that has expanded so widely via
financialization that it "produces, distributes, captures, and shapes subjectivity" (Lazzarato 2012, 32). Henri Lefebvre argues that the "quasi-logical
presupposition of an identity between mental space . . . and real space creates an abyss between the mental sphere on the one side and the physical
and social spheres on the other" (1991, 6). This has drastically collapsed
into the neoliberal logics of planning and spatial development, which have
only been pushed further toward profit maximization under austerity.

In the creative city, mental and physical space are constantly being
blurred in design materials and "projections" of completed buildings, rendered in glossy, colorful, digital life. The aspirational space of creative capital becomes real space, but its failure and its violence are enacted on the
face of the city itself and its citizens. Finance's invisible circulations do not
necessarily reveal themselves, but take on the mutated form of a neoliberal creative future rendered into space, with its glossy economic promises
foreclosed for most as its material vision is violently enacted in real time.
In condos, offices, and hotels on former ports and industrial sites, ruin lies in
the negative space occupied by capital, austerity physically crystallized in
the form of privatized "creative" spaces. This is not to suggest that we should

desire a return to previous forms of industry, but that any alternative future has already been headed off at the pass by these spatial and temporal mechanisms of the "creative fix." In the Docklands, subsumption of urban life underneath the glossy architecture, financial logics, and privileged transnational mobilities are promoted as beaming successes of financial austerity, when in fact its failure and expenditures are simply brushed out of sight, leaving a void to be filled by imaginations of common futures, but which are already circulated by private, transnational finance. NAMA administered these spectacular developments in a state of ongoing exception.

These things should worry us in more ways than rehearsed critiques of capitalism, gentrification, and culture. The ongoing tendency toward "fixes" to the systemic turbulence of global finance capital continues to amplify these unstable tides, and the political undercurrents of such responsive development logics and approaches to culture are tainted with more sinister and authoritarian circulations than now-familiar neoliberalism. Spatial fixes, creative fixes, smart and AI fixes—political dissent must respond likewise. While it is not radical enough to suggest an arts-led Benjaminian reactivation of potential within a reparative film arts practice, it is the *logic* of anti-authoritarian spatial politics through ruin that motivates artistic and other fugitive practices—such as strike, occupation, and other tactics of (counter-)planning that operate counter-logistical power or sabotage against the forces of capitalist exploitation and enclosure (see Bernes 2013; Clover 2016)—that need to be seized as a way of thinking how to see creative and media practice, within and outside of these industries, otherwise.

Struggles over space, like the occupation of Airbnb buildings by activists protesting their "colonization" of housing in Ireland (Breaknews.ie 2018) or the reactivation of a power station as a performance space, are not only fights to reclaim spaces, but are what Joshua Clover calls "circulation struggles" (2016), which respond explicitly against the movements of global capital. The disruption of capital circulation, whether in zones of finance or at the "choke points" of logistics where flow can be most impactfully blocked (Alimahomed-Wilson and Ness 2018), can all be seen as "counter-logistical" actions (Bernes 2013). This unites ongoing struggles across the world, as well as across Ireland, whether housing groups disrupting plans in Dublin, film workers advocating for better funding conditions (chapter 3), or local community members feeling abandoned struggling for the attention of capital and the state (chapter 4). Especially in film and media work, industries that have been so aggressively and fully made precarious by post-Fordist practices of systemic deregulation and the distribution of work across all aspects of the supply chain, as chapter 3 will argue, how can workers in these industries emerge as a collective force of change? How would refusal to participate in corporate projects of redevelopment, or better, active blockage of these projects, affect how planners and policymakers see those who work in these industries and live in the ruins of their "creative" spatial visions?

Chapter 3's focus on labor as a resource in media policy will deepen these analyses. By answering such questions, we are encouraged to think spatially about these industries themselves, what they can tell us about these operations of capital and the state, and where best to put our weight in resisting flagrant acts of dislocation and places of enclosure. But this is exactly why we need textured analyses of the interactions between global systems and local encounters, and their specific manifestations and choke points. The detailed approach to particular and material instances of planning and the slippages between financial, creative, and other sectors can help us come to a fuller understanding of the extent to which, in local contexts, financialization has spread into more immaterial logics of the public sphere. These creative transformations can be found in the ruin left in their wake, as much as in the spectacles designed to precede and disguise their enactment.

3 —— Waves of Austerity

Film Policy and the Infrastructural Geographies of Media Labor

While prior chapters focus on the circulatory dynamics of the media and creative industries in Ireland through built spaces and physical infrastructure, this chapter unpacks the logics of film and media policy that are *infrastructural* to the circulation of capital through the country's diverse geographies. It focuses on the disconnect between top-down systems of policy architectures and their effects on the industry and workers on the ground. Speaking formally and informally with an array of workers, practitioners, and officials across the landscape of the media industries in Ireland, from producers to tourism boosters, set drivers to policymakers, and marketing to finance executives, I was left with a picture messier than any policy or future-driven plan could ever fit into a stamped and sealed worldview. Media policy, while attempting to manage the often-diverging interests of media artists and media capital, especially in the post–financial crisis era, has frequently privileged the latter at the expense of the former. This is a result of strategic decisions to ensure the national industry's continued viability through a focus on multinational investment, as, like in other public and cultural sectors, funding receded after the crash and left policymakers to scrape together support for the industry in the absence of state money. As Geert Lovink and Ned Rossiter quip in the introduction to their edited study of creative industries discourse, "No matter how alien it appears, policy does not drift down from the heavens" (2007, 11).

To many freelance, artisanal, television, marketing, and others among the dispersed array of media workers that make up the country's media landscape, the "Irish film industry" feels as distant and inaccessible as Hollywood (see O'Brien, Arnold, and Kerrigan 2021). However, this is not to say that policy, and how it was written and who it was written for, was not a present concern among many in the sector. Screen Ireland (SI) (formerly the Irish Film Board [IFB]), the national film body, which, in the late 2010s, expanded its remit to the more expansive "screen industries" (i.e., film, TV, animation, and video games, although not the latter in Ireland as of 2020), remained present in media workers' discussions as a key overarching infrastructure for the industry, but more as a measure (and pinnacle) of meritocratic success than an everyday consideration. A career working between SI-supported projects is a goal, but not one that most saw themselves achieving anytime soon. Among interlocutors who had applied for IFB/SI funding, many felt as though there were unattainable thresholds of cultural capital required to even be considered—if you were not on the agency's radar already, funding was less than forthcoming and tended to be granted to repeat successful applicants. As a Dublin-based experimental filmmaker told me, "Once you're in, you're in," but getting to that point involved an arduous and often impossible everyday endeavor of research, networking, and luck. How could one get into this orbit without first obtaining funding? In Galway, regional funding was similarly competitive, and although I met many who applied for and received funding for projects from more regionally focused organizations like the Galway Film Centre and Údarás na Gaeltachta (ÚnG) (a state agency supporting economic and cultural development in Irish-speaking regions), these provisions were meritocratic, one-off, unpredictable, and usually would not pay the bills other than those essential to complete a project (especially if the work involved a crew).

Thus, creative media workers with aspirations of working in film tended to spend most of their time employed by contract-to-contract gigs, corporate projects, and even full-time work in other industries, while also working extra hours toward a full-time career in film. On the ground, among media workers, the film and TV industry and the promotional and industrial video industry look very similar, especially in terms of basic skills, software, who you work with, what your days look like, deadlines, and the like. Styles obviously differ, as do things like the quality of catering, but a transferable set of skills is essential. Even when looking at policy and training literature, like PricewaterhouseCoopers's series of industry manuals and guidebooks (in this case, for film financing), these materials characterize the media industries quite expansively, encompassing a wide variety of media activities from business communication to radio to film production (2016, 1; also chapter 2 of this book). Each is generally characterized, at the bottom level, by short-term work for exceptionally long hours (see Olsberg SPI 2023), although work for freelancers in marketing, for example, or seasonal work in sports media, may be better paid with more reasonable hours, depending on the quality of the project and employer.[1] Workers channel back and forth between publicly and privately supported projects and areas of these industries based on the work—from corporate videos to shoestring-budget short films and back again.

Nonetheless, despite the time constraints just-in-time and contract-to-contract gigs create for workers and the overall difficulty of securing funding for film work, among friends and interviewees, many still aspired to a career in cinema, and often spent nights, weekends, and stolen time at their day jobs working on "passion projects" building toward this goal. Many sacrificed comfortable living conditions, residing in crowded houses or overpriced apartments in Dublin and Galway. Most, in the absence of consistent funding for creative projects or qualification for limited artists' subsidies, engaged in multitudes of side hustles, smaller artistic initiatives, and other often part-time work outside of film and media. The passion and creativity involved in making a living in these industries appear operational to, as John T. Caldwell phrases it, "why we as creative subjects and cultural workers continue to *voluntarily* participate in our own apparent subjugation within the new, flexible, neoliberal economies" (2013, 106, emphasis in original). What he calls "stress aesthetics" may move us toward an understanding of how this sort of everyday precarity actually looks within these industries, in terms of the just-in-time and project-to-project hustle "culture" of the media industries on a global scale.

However, focus on ground-level flexibility must account for the complex channels that form between the state, capital, and workers, especially within a nation-state with public media policy and infrastructure. Despite reports (and common sense) citing the utility and basic need for artists' subsidies (Barton and Murphy 2020), direct subsidy seemed to be centered on "prominent creatives," whereas most of the film and media sector is characterized by low pay and overall precarious employment (Barton and Murphy 2020, 6–14). The "national industry" as studied by Irish film scholars historically (Barton 2004; Crosson 2003; McLoone 2000, 2008; Pettitt 2000; Rockett, Gibbons, and Hill 1987) is today a dispersed economy of media practitioners who interface with and contract with other sectors, which, with a few exceptions, stands in stark contrast to the promoted "national" successes of Irish films at the global box office and at the Oscars. "Media policy," especially with the collapse of "film" into streaming and domestic consumption dominated by global corporations, needs to thus navigate a widening gap between everyday media work and its modes of finance.

However, many below-the-line workers and struggling creatives do genuinely always see themselves as just a step away from their big "break" into becoming a "prominent creative," which will get them noticed by the funding agencies and big companies. As a member of the Galway-based filmmaking collective Project Spatula sloganed, "We'll keep shouting as loud as we can until someone pays attention." This meant that his organization would continue to produce work, including a microbudget and completely self-funded feature film called *Sooner or Later* (Luke Morgan, 2017), until their visibility was too great to rely on their more lucrative contract gigs that kept the creative collective afloat. While he and his team remained hopeful, such consistent difficulties in navigating funding infrastructure breed cynicism and exasperation. But in spite of consistent frustration in obtaining funding from any public sources, he requested that I check with him before publishing any findings in case the collective finally got lucky with a grant he was waiting on.[2] Despite his reasonable and justified bewilderment at the system, in the absence of security, he did not want to jeopardize his chances of success by biting the hand that might potentially feed.

Everyday relationships between precarious workers, the state, and media infrastructures are mediated through complex policy mechanisms that are developed in response to prevailing and entangled cultural, economic, and political climates. This chapter will build on the previous chapter's focus on built space in Dublin by expanding our gaze to other regions of the country, attuning ourselves to the uneven geographies of film and media policy across the diverse contexts of media industries in Galway, Limerick, and Kerry. But first, in the following section, I will outline the relationships between these media industries and the Irish state to lay the groundwork for a deeper analysis of how policy (mis)aligns with cultural labor.

In an environment of economic austerity, cultural funding is always at risk. Between 2008 and 2016, for example, funding for the IFB was cut by 44 percent. During the same period, the language and practices of Irish film policy adjusted significantly, facing the strategic imperative to not only attract capital investment in the absence of state support, but also to demonstrate the industry's potential as a tool for economic growth and development to *justify* public expenditure in the sector (see Brodie 2024a). To sustain growth, policies would have to build from existing infrastructure and "potential" (i.e., talent, skills, labor pools) by drawing in new private investment.[3] Financial incentives, industrial facilitation, and logistical coordination for global projects became the most dominant policy mechanisms over this period. In a now familiar set of arguments, made in 2009 during the financial crisis, the then-minister for arts, sport and tourism visited the set of US-Irish coproduction *Leap Year* (Anand Tucker, 2010) in Dublin and took the opportunity to extol the virtues of the recently revamped Section 481 (S481) tax scheme: "The film industry is the cornerstone of a smart and creative digital economy and with the improvements to Section 481, the Government acknowledge [*sic*] its importance at this time" (quoted in Gaiety School of Acting 2009). Over subsequent years, these caps and financial structures underwent periodic adjustment and loosening, just as the language and outward policy of IFB/SI would shift explicitly toward that of a facilitation body for inward investment (see Brodie 2024a; Murphy and O'Brien 2015).

These strategic aims did not occur in a vacuum, although the pressure of the wider business ecosystem undoubtedly affects these politics. Policy recommendations from transnational consulting firms like PricewaterhouseCoopers (responsible for a number of publicly commissioned reports on the Irish "audiovisual sector") (i.e., Department for Arts, Culture, and the Gaeltacht 2011; PricewaterhouseCoopers 2020) worked to reduce the film industry to profit-making and return on investment and the value of "content" within a competitive transnational digital marketplace (PricewaterhouseCoopers 2016, 3). In 2011, a steering committee for the "audiovisual industries" released an influential report, modeled after a 2008 PricewaterhouseCoopers analysis of the Irish "audiovisual sector," entitled *Creative Capital: Building Ireland's Audiovisual Creative Economy* (Department for Arts, Culture, and the Gaeltacht 2011). This report, beginning with a quote from creative industries guru Richard Florida, outlined a series of policy recommendations to build capacity and capitalize on the economic growth potential of the sector. While direct references to the ongoing economic recession are limited and claim that the audiovisual sector remained "resilient" despite the downturn (2011, 32), one can infer the degree to which public incentives would have to be justified via economic growth, privatization, and foreign relations. Asserting that the sector was sustained on FDI and that "creative and cultural industries are a powerful motor for jobs, growth, export earnings, cultural diversity and social inclusion," the report recommends "specific policies that will equip the industry to successfully enter the next phase of its growth from a predominantly domestic platform into international markets" (i–ii).

The focus on global success as the essential guiding force of industrial viability also demonstrates the wider shift toward creative industry logics surrounding the place of production within creative economy "growth" and discussions around digital "convergence" in global film and media (Jenkins 2006). This practical and rhetorical shift followed by Ireland's public bodies is evident from simple promotional materials and the IFB's rebranding as "Screen Ireland" in 2018, with a revamped website advertising the country's infrastructure, tax breaks, and available creative labor as reasons to invest and produce content. In expanding the scope of their operations to projects outside of the typical banners of feature, documentary, and short films in order to incorporate a more wide-ranging idea of the "screen industries," they have plugged into a more diverse set of creative industry sectors and corporate development strategies.

However, while these new mechanisms for directing public resources into multinational profits are the result of austerity logics and global macroeconomic shifts since the financial crisis, the politics of privatization are deeply and historically imbricated in the policy language of the Irish state and how it sees media as a social and a spatial force for economic development. These FDI-driven logics, however transformed under "convergence" and the tech-driven shape of film and media in the 2010s, have undoubtedly been the guiding hand in the background of Irish film policy for some time, if not in the arts more widely. This crux is both at an industrial scale—in terms of incentives for the "screen" and audiovisual industries—and also enacted through visual media in a more traditional, representational sense. Cultural theorist Justin Carville contends that Irish visual culture has frequently been characterized by an "absence," as images of Ireland have historically either come from outside of the country or been directed to a touristic audience (2013, 17–18). This idea has implications not only for

"Irish visual culture" broadly, but for how we study Ireland, its audiovisual production industries, and the visual culture produced of and by the state, its public institutions, and its private partners, as suggested in chapter 2. National culture and global capital have historically been entangled within "Brand Ireland" for decades at this point. As Luke Gibbons points out, the visual culture, ideology, and propaganda of central government planning (with bodies such as the IDA) were important in the development and naturalization of the shifts toward globalization (1996, 86–88), and these logics were both intensive and extensive in scope and action.

State infrastructure around the film and media industries has tended to be top-down projects, focusing on large-scale activities as assurances of viability. Most film productions before the first IFB, established in 1980, were inward productions, often facilitated by the intermittently national-ized Ardmore Studios in Wicklow, which was specifically established in 1958 for Hollywood-style productions during Lemass's tenure as minister for industry and commerce (see Barton 2004). Films like *The Spy Who Came in from the Cold* (Martin Ritt, 1965) and *Zardoz* (John Boorman, 1974) were shot at the studios and in surrounding locales in County Wicklow, and much of the nation's film infrastructure and labor resources are still centered in Wicklow and nearby Dublin. However, pre-IFB production in Ireland was most prominent and recognizable globally when it featured rural, idyllic locales, with films like *The Quiet Man* (John Ford, 1952) and *Ryan's Daughter* (David Lean, 1970), shot and set in the west of Ireland, receiving international attention. Such films, part of much longer cultural visions of the "wild Irish" coming from England and the United States, paint Ireland as a place of backward and regressive people and politics, whether positively or negatively romanticized as a place of quaint simplicity or vio-lence, romance or "boggy mediocrity" (O'Toole 1997, 19).

The global "imagination" of Irish culture and landscapes expressed through such films, the "Emerald Isle" trope, has been a straw person within dominant studies of Ireland as a "national cinema," an action that frames the domestic industry in opposition to the peddlers of trite, stereotyped, and commodified representations (Barton 2004; Crosson 2003; McLoone 2000, 2008; Pettitt 2000; Rockett, Gibbons, and Hill 1987). As much as cultural analyses are crucial, the stakes of these audiovisual representa-tions can be explained and understood in the powerful force they have exerted over the political, economic, and infrastructural circumstances of Irish film and media production and policy. How do the spatial and terri-torial practices of this "national" film culture, incentivized by the state and nourished by on-the-ground workers, artists, and "creatives," map onto the socioeconomic shifts brought about by the financial crisis? And especially the transformed structures of dependency, as can be seen and experienced in the current FDI-driven facilitation model of the "recovery"? And finally, how are those workers treated and managed across spatial and infrastructural considerations within these logics?

The use and development of Irish space for media production since the crisis demonstrates increasingly common slippages between processes of financial speculation and cultural production. In a suitable microcosm of the post-crisis economy, a state-of-the-art film facility in Limerick, Troy Studios, was built on a brownfield former Dell manufacturing site. Dell had moved its manufacturing operations elsewhere and shifted its Irish focus to enterprise services in Limerick and Dublin (see Collins and Grimes 2011), leaving the industrial ruin where there would have once been a supply of computer components for export.[4] At the same time, the national film and media industries needed more studio production space for foreign productions. Media and tech industries hold a key role in spatial transformations as property finance migrates to new deregulated sectors for growth (see chapter 2). In a meeting over coffee in Galway in 2018, the CEO of Troy Studios at the time told me that the company had acquired the site on a twenty-year lease from Limerick City Council, and that the primary attraction was the existing structure, as building studios from scratch is not as economically viable as repurposing existing industrial-scale structures. The studio has expanded its soundstages to become the largest in Ireland, with a total of over 100,000 square feet of available filming space, in addition to surrounding supports and facilities (Hamilton 2019). The studios, located in the National Technology Park adjacent to the University of Limerick, were intended to act as a training ground for workers and technicians in the area. Thus, in the late 2010s, groundwork was being laid for a future film infrastructure in Limerick, centered on the studio: Like the market more broadly, Troy, as its CEO paraphrased, would bring a "rising tide" to "lift all boats."

Converging headlines pairing industrial ruins, regenerative planning, economic development, and media production concerns were a common occurrence in the national and trade presses post–financial crisis. Apart from publicized proposals for film studios in ghost estates (Dowling 2011), Dublin Bay Studios, a joint venture between film and TV producer and Windmill Lane founder James Morris and Parallel Films producer Alan Moloney, attempted to secure a brownfield former manufacturing site in the Poolbeg West SDZ that eventually was awarded to a mixed-use property development. The proposal was widely supported by the government (and U2 frontman Bono) (IFTN 2017), but opposed by the Dublin Port Company, which retained final say over the zone's use (see chapter 2).[5] These zoning directives are mobilized on an uneven basis depending on stakeholders. While Dublin Bay Studios was competing for the Poolbeg West SDZ in early 2017, Ardmore Studios in County Wicklow, long the home base of film offshoring in Ireland, was fighting to avoid demolition and redevelopment as a private housing estate. A revolt among media workers led to pleas for its nationalization as a "going concern," and the government recognized its zoning as a "film-only zone," with an initial potential for a workers' buyout of the property (M. O'Halloran 2017). However, in March 2018, the studio was bought out from the previous shareholders (including semi-state Enterprise Ireland [EI], which supports Irish business interests abroad) by private equity firm Olcott Entertainment,[6] which pledged to retain and expand its function as a production site for film, TV, and video games (Bodkin 2018). In this case, the special zoning directives of the Irish government saved local industry about to be foreclosed (to be contained within private housing) through full privatization (with state protection) of the country's first film studio.

These complex forms of mediation between the state and the private media sector, frequently implanted through spatial development and in relation to its management of global investment for industrial activity, demonstrate the infrastructural role of policy in the construction and management of media activities as a viable tool for economic development. In securing Irish space for global media productions, Irish policy organizations become facilitators of multinational capital, while also performing the necessary role of securing employment, training, and other opportunities for local crews and economies (see Brodie 2024a). However, it is within these apparently necessary compromises—wherein a "national" film culture becomes a product of secondary training and skills development offered by spillover from global production—that the insecurity of such practices, especially in a context of financial crisis, becomes more evident.[7] Moreover, the very infrastructure of these organizations must change in order to operate within such a competitive global economic landscape, directing state resources toward wooing international investors rather than directly investing in direct public supports.

Moreover, finance capital is uniquely primed to circulate through an environment so often thought about in terms of cultural capital and national self-determination. The embattled "public" purview of cultural enterprise often allows financial logics to operate largely without criticism, under the cover of apparent public interests and benefits to the population.[8] As argued in chapter 1, the financialization of media is a central tenet of media globalization and references the permeation of financial logics into spheres apparently disconnected from the financial industry (see Haiven 2014). As I have argued elsewhere in an article on the specificities of this scheme, "Overlaps between financialization and supply chains occur across the diverse geographies of media: tax incentives attract foreign media capital, which extracts value and funnels profits through financial instruments back to productive film centres like the US, UK, and India. Until 2015, financial instruments, such as special purpose vehicles (SPVs), short-term companies put in place to accommodate FDI and avail of tax benefits, acted as conduits through which media finance circulated the space of Ireland" (Brodie 2024a).[9] The 2011 *Creative Capital* report makes this financialized media strategy more explicit, arguing, amid a financial crisis driven by irresponsible banking, that greater engagement with the banking sector would be essential. "Angel investors" and "venture capital" should be courted, and "strong relationships with the banking sector are important and should be a key function of enterprise development" (Department for Arts, Culture, and the Gaeltacht 2011, 7). As I argue,

> The Section 481 tax scheme is designed to put these financial conditions in place. Previously, a variety of actors could invest in a media property as part of speculative investment portfolios. However, in 2015, the scheme transitioned from, as Denis Murphy and Maria O'Brien describe, "investor-led" to "exchequer-led" (2015), meaning global investor risk was replaced by Irish taxpayer money. It ostensibly ensures that companies do not take liberties with public benefits, and that an Irish workforce is hired, for example, by mandating that all FDI must partner with an Irish production company or establish a long-term subsidiary in Ireland. However, large-scale productions can buy companies to still avail of the benefit depending on how they are able to justify their spend, employing whatever workforce they need to finish the project. (Brodie 2024a)

Broadly, such mechanisms, while so naturalized as to appear commonsensical, foreground the tentative promises of social benefits as a justification for redirecting public investment into what is ultimately risky private enterprise. By these logics, securing investment for an underfunded sector depends on mediating and mitigating investor risk via public funding.

But while these logics are familiar within the geography of financial globalization and spatial planning in Ireland more broadly, as chapters 1 and 2 demonstrate, they are newly explicit within culturally oriented policy, especially since the financial crisis. The geographies of such financialization reflect and perform in tandem with global media supply chains and their coordination and management across spaces. SI has reorganized itself as an agency designed for facilitation of and in deference to FDI and the logistical enablement of its operations. In a section after its Screen Ireland rebrand called "Why Film in Ireland?," SI's website read: "Ireland is a world-class dynamic filming location. We provide logistical on the ground support for film and television projects filming on location in Ireland" (Fís Éireann/ Screen Ireland 2020). "Why Ireland?" is a question local and national economies must answer to draw in foreign capital and create jobs and revenue in the short term, justifying the existence of the public agency by inserting it within a variety of buzzy, circulatory dynamics key to the functioning of a successful creative economy. In a section called "Filming in Ireland," it stated: "Ireland is a dynamic and world-class location for international production. Our excellent crew base and technical infrastructure combined with our breathtaking scenery and generous tax credit make Ireland a central hub for TV, feature film, and animation" (Fís Éireann/Screen Ireland 2020). All text was positioned across backdrops of existing productions and breathtaking landscapes. As chapter 1 argued and chapter 4 will develop in terms of the Irish "business climate," industry and scenery have coalesced in the visual and promotional culture of industrial development projects in Ireland since the earliest stages of liberalization. In SI's promotion of the industry, landscapes, financial incentives, labor pools, and technical infrastructure formed an apparently thriving film industry machinery despite an ongoing financial crisis, requiring only periodic influxes of transnational capital to fire into action.

Policy and regulatory agencies, as we know, exist across borders as much as within them. Even ostensibly "national" bodies like the MPA (formerly MPAA) in the United States work, in overtly imperialist ways, to exert influence abroad via market-control mechanisms like intellectual property enforcement. But this "soft power" extends to film commissions and cultural semi-state organizations as well. Thinking in this way, the uncanny and place-agnostic operations of state, semi-state, and even corporate soft power make sense: that, at the time of my research, SI held offices in Los Angeles; that Irish film studios sent representatives to Los Angeles and London to promote Irish production facilities; that semi-state body Culture Ireland sponsored and funded Irish cultural activity abroad at international festivals, events, and conferences; and that the supranational body Creative Europe held two "desks" in Ireland, one in Dublin and one in Galway, the former of which shared a building with SI's central offices. The intensive and extensive operations of corporate media policies and industries demonstrate that "soft power," which relies on cultural, economic, and civil society processes to influence international movements and relations (such as investment patterns) (Nye 1990), has a concrete geography, working across territories and scales of governance.[10] Its politics represent multilateral movements and agreements across territories and actors, where the power of capital and sovereignties is experienced and enacted differently depending on where you are standing (and/or working). With this national and cross-border operation of policy in mind, we can come to a greater understanding of how nation-state organizations and policies, far from acting only in response to transnational forces and pressures, are entangled with the very same flows of capital and influence that they are designed to regulate and keep in check. In the case of the film and media sector, this is done by cultivating a cultural milieu and workforce within their borders designed to support global production.

Pertinent to Sandro Mezzadra and Brett Neilson's articulation of the role of the state in managing crisis as intervening to facilitate rather than regulate capital (2019, 49), cultural policy post–financial crisis in Ireland has been a site of contestations around austerity and the use of space (see Guinan 2016). During the crisis and subsequent austerity measures, funding for cultural production in Ireland receded, to be replaced by mechanisms for "cultural enterprise" and an expanded tax break (S481) for media production and postproduction on Irish soil (between 32 and 37 percent, the graduated rate designed to draw investment into non-Dublin and Cork regional production).[11] With this understanding in place, it is easy to see the circular thinking that drives the transnational logics of media production— the media sector has to make money domestically to provide jobs and ensure a constant workforce; which means that foreign productions have to be attracted; which will eventually build native talent and enterprise; which will then create jobs. Infrastructure will take shape from the "ripple" and "spillover" effects of this cycle, which will crystallize into a more robust and secure environment for workers and cultural production. This is a policy logic that can be transplanted across sectors. But the ongoing presence and influence of the territorial expansion of multinational corporate media operations, intensive and extensive, will never release film and media policy and its public remit from the precarity of this double bind. Even the most resilient infrastructures built out of this cycle will still be tied to the roving fortunes of global media capital driven by competition between territories and shifting transnational policy and regulatory conditions (see Mayer 2017). Mezzadra and Neilson's framework provides us insight into the ways that "interdependence" drives state responses that have just as much effect on territories as do the nomadic operations of global capital, despite the "global" power structures defining the operations of multinationals across territories.

It is thus instructive to think about why states continue to privilege these economies despite the obvious subjection to unstable and often exploitative global industrial formations. As many argue (Curtin 2016; Mayer 2011, 2017; Miller et al. 2001), film production has participated in a global race to the bottom in terms of various service points along its supply chains. While cinema is a "highly skilled" medium by governing job standards (as though there are not a multitude and layering of jobs and tasks in the media production ecosystem), there are other ways to cut corners than simply moving to deskilled, underpaid environments, and the structure of uneven development still dictates that "creative" workers in the Global South are paid less than their northern counterparts (see Mayer 2011). As Michael Curtin elaborates,

> Media and markets have long conspired to privilege some places over others. These inequities have prompted some governments to persistently assert their cultural authority and political sovereignty by intervening in the realm of popular communication, the result of which is that political capital sometimes counts as much as, if not more than, economic or creative capital. In extreme cases, media have at times become instruments of the state or of an autocratic elite. At their very best, however, public media institutions and enlightened media policies have provided the means by which common legacies have been maintained, policies debated, and futures imagined. (2016, 675)

This proliferation of short-term, project-based production has expanded territorially to encompass entire nations and regions bidding against one another for these investments and what they offer to the "public" service of a media industry. Cultural diplomacy becomes not a struggle or even a negotiation for cultural territory and protection, but a competition for private transnational capital, lauded for its ability to provide an influx of capital into otherwise struggling sectors during times of, for example, austerity.

Keeping in mind how this transnational order reshapes the relationship between the nation-state and cultural production, we can thus see how viewing government and transnational policy as *infrastructural* to the contemporary arrangement of Irish media industries can help us build a better idea of its everyday functionalities for media workers. How do sovereign discussions and diplomatic arrangements that respond to transnational movements affect (or appear not to affect) those working in "national" film and media industries? And what happens to ideas of "public" and/or "cultural ownership" when these cease to be the primary source of value in policy discourse? Or, more simply, do transnational modes of production *actually* sustain Irish media workers with meaningful and robust infrastructures?

The material production practices occurring in transnational and global preproduction deals, production funding, and production and postproduction cultures, thus present a twofold problem for the study of Ireland's film and media sector. On the one hand, there is a need to understand wider infrastructures of media production and circulation at a transnational scale through the geographical contexts through which these infrastructures must pass. These vectors of power and influence *do* land on communities and workers in ways that force scholars and advocates to always keep eyes on them, lest they shift into even more troublesome identitarian territory.

However, this relocalization risks, on the other hand, contributing to conservative focus on the "cultural identity" of Irish film and media despite the realities of transnational production and circulation as the defining characteristics of the contemporary media landscape, especially through the manufacturing of the creative industries via strategic activation of cultural heritage and arts traditions. These tensions are felt across media and cultural policy as well as scholarship on it. As Huw David Jones asserts in a study of UK/EU coproduction agreements, "even though most coproductions are financially driven, the logistics of co-production inevitably have consequences for a film's cultural identity. Official co-production treaties often specify that each co-production partner must have a creative input in the film which is proportional to their financial investment, even if this is not always appropriate to the film's narrative" (2016, 5). This ongoing focus on cultural identity in policy—espoused in policy arithmetic by "cultural tests" to determine not just funding but also the advertised identity of a film based on majority and minority coproducing partners—is an economistic mechanism by which space and labor frictions are eliminated within co- and offshored productions. Apart from much-maligned erasure of cultural specificity, such policy aligns identity with a register of capital investment and above-the-line talent. It irons out the granularity of diversity within a much wider film and media industry supposedly represented in the credits of globally and coproduced products. Studies of the actually existing (often lack of) diversity in the Irish film and media industry notwithstanding (see Asava 2013; Kerr 2000; Kerr and Preston 2001; Kerrigan and O'Brien 2020; O'Brien et al. 2022), the transnational policy arithmetic of Irish state and EU agencies is an imaginative mathematics that ignores conditions of precarity and exclusion at the heart of film and media production economies, which maps onto other forms of precarity and discrimination in the workforce (O'Brien, Arnold, and Kerrigan 2021). So, while studies of small national cinemas tend to focus on policy and output, this chapter supplants focus on mostly above-the-line cultural identity, administered by "cultural tests," funding, and public discourse, to trace conditions of media production and the implications of such policy math for workers on the ground, especially those said to thrive on the "ripple effects" provided by large-scale FDI in media.

In the remainder of the chapter, I thus foreground the testimonies of an array of film, media, and cultural workers across sites in Ireland, collected formally and informally between June 2017 and May 2019. I paint a very different picture than the one promoted by official state bodies and media. That is, I articulate how workers relate to public organizations, whether with gratitude, indifference, or antagonism, in relation to their perceived conditions of media and cultural industries, work, and funding.

Social Infrastructures and Regional Specificities

Walking around Galway city in the 2010s, the seat of County Galway and the fourth largest city in the Republic of Ireland, one would be struck by the presence of the arts, with the main paths through the city center populated by buskers, performers, and the revelry of its compact tourist center. Pop into any pub or event space in the evening and there would invariably be some sort of live event, from cover gigs to DJ sets, to performances, poetry, and film screenings. Seasonal festivals regularly enliven the cultural scene, and many are employed or rely on revenue generated by large-scale yearly events like the Galway Film Fleadh and the Galway Arts Festival, among many others. Work bleeds into leisure, and days off are often spent in one of the city's dozens of pubs drinking pints and chatting culture, arts, and politics. Many of my friends used to call Galway the "graveyard of ambition," a general feeling within which I often found myself completely seduced during visits. Much of my general insight also comes from extensive time spent having pints and discussing the arts scene in Galway and across the country, and thus, like much of the fieldwork for this chapter, may appear experiential in nature.[12]

That said, to reduce Galway's arts economy to that of a bohemian party town is obviously to diminish the social role that these sorts of environments play within broader regional, national, and transnational economies, in addition to the socioeconomic character of such work and the realities of living in a "creative city" in post–financial crisis Ireland.[13] After all, Galway was a UNESCO-designated "Creative City," which implemented the "Galway City of Film" initiative beginning in 2014. The buzz created by such policies and the economies they are meant to support do not spare the space from the more familiar undersides of the uneven effects of the "creative city" described in chapter 2. Among film and arts professionals in Galway, I met dozens of artists, performers, and workers, most of whom worked on contract-to-contract, freelance, or otherwise nonpermanent, flexible, or precarious bases in their fields. Buskers performed next to rough sleepers. Artists and media workers struggled to make rent. College students lodged with families subsidizing mortgage payments with sublets. Many young people contemplated leaving the country on short-term visas for work opportunities and experience, and those who returned found the job environment inhospitable to those who had not put in their time networking and ingratiating themselves into the local or national scene. The socioeconomic underbelly of such a scene remains the persistent reality of life in post–financial crisis Ireland.

Especially in the creative and cultural industries, many people went abroad during the financial crisis to look for work due to lack of funding and opportunities.[14] During the recovery, émigrés started to trickle back into the national workforce. However, despite the promises of opportunities and skills development abroad, some had come back and found that they had lost a step in the local industry that is supposedly there to build a more robust on-the-ground labor pool and technical infrastructure. For example, a Gaeilgeoir founder and director of a Spiddal-based Irish-language theater troupe and TV production company I spoke with had spent time abroad in the United States working when he was younger, but emphasized that the struggle for funding and work in the Irish industry was a long game. To him, you needed to know the community, meet funders in person, repeat meetings, and build relationships in order to maintain a publicly funded media and arts practice. As it turns out, time spent on the ground and in the community was more important for subsistence work than flashy foreign credentials.

In a conversation with two filmmakers at a small café in Galway one afternoon, both characterized the professional media environment of Galway as one of festivals, conversation, networking, and deal making, whereas the majority of the country's major media production and postproduction still occurred (or was centered) in the Dublin region. This was despite regional uplifts added to the s481 tax relief scheme. As several told me, s481 had less effect on local productions outside of Dublin, tending to subsidize the higher scale of production and postproduction, and even suggested that it served to draw workers *from* Dublin rather than necessarily employ local talent. Other regional initiatives and subsidies included the Western Region Audiovisual Producer's (WRAP) Fund, a subsidy by the Western Development Commission, ÚnG, local county and city authorities, and the only remaining regional film center (as of 2019), the Galway Film Centre. The fund would provide up to €200,000 for any "feature film, television drama, animation or game that undertakes a significant portion of its production (including post production) in the Region (Clare, Donegal, Galway, Mayo, Roscommon and Sligo) and can demonstrate a strong prospect of generating a financial return on our investment" (WRAP Fund n.d.). The fund also cited local infrastructure—production companies, studios, locations, and transport—suggesting that it was open to attracting inward productions.

But these types of funds could not be counted on by most media workers, nor could the number of productions supported by them, to provide sustainable work. Many of the most successful media professionals brandished a self-promotional entrepreneurialism, forged by necessity. The small-scale and intimate social environment of the arts also meant that projects dwelled in association with other art forms, organizations, or companies. A member of Project Spatula lamented the difficulties of securing funding and said that he and his team had established themselves from the ground up, based on what limited work was available around them and an entrepreneurial and innovative ethos. The collective made great efforts to pay all their staff for every project (including passion projects), which counted among its members theater professionals, writers, nonprofessionals, as well as film and media technicians, but it was difficult to do so relying only on private contracts. The Spiddal-based children's TV and theater producer I spoke with had started his company after establishing himself and working his way through the public TV system. Filmmaker Liza Bolton was heading up a mini film festival called "Shot by the Sea," funded by the village of Salthill and the Small Towns Big Ideas project of the Galway 2020 European Capital of Culture initiative, which at the time was undergoing a period of turbulence in its management and personnel, meaning that many felt unsure of its actual direction. But while Shot by the Sea's credentials and ambitions appeared to transcend its local scale, what Liza described to me insisted otherwise. What she wanted out of the project was a community-centered initiative to celebrate the local coastal landscape through short films, and the funding from Salthill—a small coastal community and resort town just outside of Galway city—reflected an arrangement mutually beneficial to both creative expression and community enterprise.

Little Cinema Galway was a similar initiative deeply grounded within the city's film scene, hosting a monthly exhibition of local short films at the Róisín Dubh bar and a yearly forty-eight-hour film competition. Such community-oriented endeavors existed in tandem, tension, and independently from broader schemes. One member of a radical theater collective was also a freelance media worker and was involved in documentation of an Irish-language hub for teaching high-tech skills through Irish. He said that top-down projects like Galway 2020 (through the European Capital of Culture initiative) tried to artificially engineer the creative industries, whereas he was actively trying to collectivize arts and media workers to forge alternative infrastructures (like the theater collective). So, while initiatives and institutions like the S481, WRAP Fund, and Galway 2020 reflected the perceived macroeconomic necessities of creative work in a traditionally rural region, incentivized at the public level, smaller-scale projects more directly involving local communities were cobbled together from whatever else would be available at any given time, whether small funds, community participation, or individual expenses.

Thus, the crucial elements of film and media in Galway were not just the formal infrastructure put in place by the state or civil society arts organizations. Rather, just as, if not more, important were the informal exchanges that occurred in the community, within arts events, meetings in pubs and coffee shops, and the everyday hustle of getting by, making work, and moving up in the national film and media industry. The coexistence of networking and entrepreneurialism, often by necessity, with community-oriented practices, more explicitly political in their aims, reflected a broader tension between industry and arts practices within the media industries. As the two filmmakers said during our long, relaxed conversation over the course of an afternoon in Galway, established institutions like the Galway Film Centre and the yearly Galway Film Fleadh remained somewhat apart from the arts community that sustained the city's bustling film and media scene, even though many worked for them, sent work, or counted on them for networking. While they obviously and necessarily overlapped, most work was made and contacts created through existing and everyday social networks in the local industry.

Even the lottery of achieving public funding could not be gamed properly if you were not socializing with the right people. If you wanted to avoid bumping into a friend or colleague, you had to avoid the city center, and even then you needed to be careful—a friend referred to a look over both shoulders, practiced by most before gossiping or shit-talking, as "the Galway shuffle." In the arts community, everyone knew everyone, and anyone could be anywhere. I saw countless gigs awarded over an evening pint, where last-minute short staffing or serendipitous connections led to collaboration or, more consistently, a much-needed paycheck. Folks sometimes sacrificed consistent or permanent employment for the "freedom" of freelancing to work on a string of projects, which may be more interesting and nourishing, even if precarious. The "hustle" of everyday media work was a direct response to what is available and viable in a competitive but also intimate local film and media culture. In such environments, you meet few people who have ever been awarded substantial state support directly for their work, unless of course you count the dole.

Ripples in the (South)West

Apparently far from the everyday, ground-level hustle of film and media work in Galway, Troy Studios, a global film studio built on the brownfield site of a former Dell manufacturing plant (see figure 3.1), has been at the center of Limerick's top-down project to cultivate a local industry. The studio was designed to attract large-scale foreign productions and experienced workers locally and from elsewhere in the country to relocate for long-term work and to use s481 projects to train a local labor force. However, these sorts of projects and their "ripple effects" are subject to the fragile whims of the global media marketplace when tied to the success of anchor clients, even if effectively managing to attract investment and new talent to the area in the short term. Using a similar strategy, Ashford Studios in Wicklow, established in 2012, lucked out with its first client, History Channel's period epic *Vikings* (2013–20) having such a long and successful run, employing hundreds in the region for a period of several years. However, *Nightflyers* (2018), Netflix's George R. R. Martin–adapted sci-fi television series and Troy's first major production, was not so fortunate. Pitched as a potential workhorse for the local industry, its abrupt cancellation immediately after its poorly received release left the studio without a constant tenant for the near future, leaving the proposition for local industry and workforce built on the back of such place-averse foreign funding tenuous at best.[15] In spite of the growth in such studio infrastructure in Ireland, a thinly coded message from a Disney executive at a public talk at the Dublin International Film Festival in 2019 felt meaningful: He said that the skills were getting there across the country, but the technical infrastructure—primarily studio space—was not there yet for the country to become a real global media powerhouse (Bailey 2019).

However, while much attention on Limerick's media industry was focused on Troy and its facilitation of large s481 projects in the late 2010s, the cultural environment of Limerick was centered on its vibrant arts scene. In a jaunt through the city in 2019, I met with a variety of artists and media workers, getting a limited but revealing insight into a cross section of the cultural activity in a city that, according to several of my interlocutors, had never really experienced the "beneficial" growth effects of the Celtic Tiger boom, thus leaving it in a unique position post–financial crisis. That unfortunately had not spared the city from the housing crisis and spatial inequalities that intensified during the recovery. An artist and graduate student who had recently moved to Limerick from Dublin balanced her arts practice with tenant organizing. Her arts practice, she told me, felt completely inadequate to actually address the depths of Ireland's everyday and crisis-level problems with housing and inequality.[16]

This artist-researcher put me in touch with several colleagues and institutions, including the Ormston House, an arts center established at the height of the crisis in 2011 as a cultural resource center and developed further as part of Limerick City of Culture, the last remaining artists' space in the city center operating from this period. While started as a city-funded initiative, it took on a life of its own, even buying its current space out from under the speculative gazes of US vulture funds guided by NAMA. Unlike Troy, which had been established with a particular infrastructural purpose in conversation with government policies, Ormston House, as an ad hoc arts studio and performance space in the ruins of the former Dell factory

3.1 Troy Studios, Limerick, in winter 2019. The studio building, surrounded by a security fence, is too big to capture fully in frame from any available vantage point. Photo by Patrick Brodie.

during and after the Creative Limerick program, was an infrastructure that responded directly to community needs.[17] As the company's mission statement read, "Our core question is: *How can we support artists better?*" (Ormston House n.d., emphasis in original). The center prided itself on embeddedness in the Limerick community, especially as a space for diverse audiences and participants cutting across the city's stark socioeconomic divides. The codirector at the time was certain a space like this (and a cultural scene like that in Limerick) could not exist in Dublin or elsewhere in the country. She said that Limerick was uniquely sized, with its own micro-history of economic development and social stratification, and any true community arts space needed to behave accordingly based on community requests. Their projects, even those geared toward professionalization and arts training (two of their main remits), were grounded in community wisdom and public interfacing.

Creative industries discourses, on the other hand, often marginalize the critical role that such community knowledge practices provide, unless employed toward explicitly economic goals. One sector within which this is particularly pertinent is tourism, and the service industries in general, which are situated uneasily within cultural policy and its economistic measurements (except in the official "heritage" sector). The landscape and people's relationships to it are central to how governance is unraveled, especially in more rural spaces. In fact, Ireland's industrial policy enfolds the "natural" landscape more often than not (see chapter 4), even in cases of natural heritage. For an example: From 2014 to 2016, Disney and Lucasfilm secured permission—under controversial circumstances—to film scenes for *Star Wars: The Force Awakens* (dir. J. J. Abrams, 2015) and *The Last Jedi* (dir. Rian Johnson, 2017) on UNESCO heritage site Skellig Michael and later surrounding locations like Ceann Sibéal in west Kerry. Although the Dingle Peninsula has long been visited for sweeping vistas made famous in *Ryan's Daughter* (dir. David Lean, 1970), other sites in west Kerry like Ballyferriter, Portmagee, and Skellig Michael have been advertised for *Star Wars* tourism both locally and on the corporate website of Wild Atlantic Way (a drive tourism route encompassing the western coastline of Ireland) since before the release of the first film in the new trilogy (Episodes VII–IX) (see figures 3.2 and 3.3). As Ruth Barton observes about the dubious tactics used by the *Star Wars* production and their partners in the state (including the Irish Film Board) to film on the UNESCO World Heritage Site Skellig Michael, this sacrificing of public heritage at the altar of profit is widespread across Irish economic development (Barton 2019, 307; see also O'Toole

3.2 "The Last Beehive Huts" in Dingle in winter 2019. A farmer had re-created ancient monastic structures featured prominently in the *Star Wars* franchise, charging €3 for admission. Photo by Patrick Brodie.

3.3 A placard locals put up near Ballyferriter to memorialize the *Star Wars* shoot that occurred at Ceann Sibéal, the prominent cliffhead in the background, winter 2019. Photo by Patrick Brodie.

2015). The "devil's bargain" (see Urry and Larsen 2011) of such facilitation applies to both production and tourism, as local hands are expected to enable and reproduce a touristic view of space that may have completely distinct local meanings or values.[18]

Key to arguments about the sustainability of film and media industries when justifying public subsidy are ripple or spillover effects of tourism. Crucial for this argument, especially around the economic sustainability of media-driven tourism (as a justification for foreign productions), is that its justification is initially production-driven. Based on industry reports detailing strategic recommendations for the Irish audiovisual sector (Crowe Horwath 2017; Department for Arts, Culture, and the Gaeltacht 2011; Irish Film Board 2016; Olsberg SPI with Nordicity 2017), the 2017 Audiovisual Action Plan published in conjunction with the Creative Ireland program recognizes the already-existing synergy of film production and tourism, and that film incentives draw future tourist traffic: "The Section 481 tax incentive will continue to contribute to the development and sustainability of the Irish screen industry, supporting jobs in the domestic economy, a strategic cultural industry and the tourism sector in Ireland" (Department of Culture, Heritage and the Gaeltacht 2017, 10). However, less prevalent throughout the policy literature are concrete measurements of these benefits of tourism, which are found in abundance regarding investment, tax revenue, and jobs in production. The Olsberg SPI with Nordicity report argues that "whilst this analysis indicates that Section 481 is breakeven on a [fiscal net benefit] basis, it is important to keep in mind that the scope of this research has not quantified some of the spillover effects—particularly screen tourism—which can often add a significant amount to economic contribution and resulting tax revenue. For this reason, the breakeven result should be viewed as conservative" (2017, 27). In two cases in the report, the estimates of this benefit are viewed as "conservative," while tourism is cited as a predominant contributor without accompanying figures. In a more recent report from PricewaterhouseCoopers, "screen tourism" is given an entire chapter, with frequent vague references (visual and written) to the "success" of *Star Wars*, even though most of the success stories come from case studies outside of the country (thus encouraging development in the sector while not outlining many existing benefits) (2020, 35–40).

In my research, it was difficult (if not impossible) to measure the spillover success of tourism.[19] Most I spoke with had yet to see the results beyond dedicated *Star Wars* fans and unplanned encounters by tourists already there for the heritage and scenery. In 2019, a Dingle-based tourism operator who I spoke with estimated that about 2 percent of his inquiries were *Star Wars*–related, despite a dedicated page he made for his website roughly a year before my visit (Dingle Slea Head Tours n.d.). He took me on this version of the tour, which he admitted was essentially the same as his other Dingle day tours, because there was not much left to see from the production at the Ceann Sibéal or other minor sites around the peninsula due to environmental agreements to leave no permanent trace on the landscape. Sites across the peninsula had informal signs branded with *Star Wars* insignia indicating where production had occurred, including competing claims to cliffside locations between landowners. However, as the site was a stand-in for Skellig Michael, most tourist traffic had been directed there, a place with a robust existing clientele due to the UNESCO-recognized natural and monastic heritage of the island. The manager of the Ceann Sibéal Hotel in Ballyferriter gave a small estimate similar to the tour driver, saying she had received about a dozen guests expressly interested in *Star Wars* between 2017 and 2019. While there was community pride and enthusiasm for the production, as seen in wide commemoration and vernacular merchandising of the brand between schoolchildren, craftspeople, and small merchandise companies (which sometimes brought them into interesting conflict with Disney's copyright rules), widespread tourism of the kind seen in other locations had yet to follow. Local providers in Ballyferriter could easily recall most of their few *Star Wars* inquiries in detail, standing out among the thousands of ordinary tourist visitors they received each year.

Spillover figures aside, there are more unexpected synergies between film and tourism infrastructure, particularly below the line. For example, drivers and boat operators who circulate tourists around to sites in the Dingle Peninsula and Skellig Ring (where trips to Skellig Michael usually departed from) also drove crew, equipment, talent, and provided other services during *Star Wars* production, including the purveyor of my own Ballyferriter tour, who helped recruit other local drivers. Despite the construction crews and materials largely being brought in on lorries from Dublin with crew mostly from there and the UK, tradespeople and manual laborers across the peninsula reportedly were employed by the shoot, along with those in service industries. Many "young lads" in manual labor services around Portmagee helped carry materials up Skellig Michael, working industry-standard long days to load crew and materials in and out. The skills transfer in these jobs moved seamlessly between transportation and labor in both film production and tourism, emphasizing the importance of blue-collar and below-the-line work in creative and cultural industries.

These types of work are not accounted for within cultural policy, a surplus labor pool and "resource" taken for granted. A boat operator I met in Portmagee, who had a license to ferry tourists over the treacherous patch of sea out to Skellig Michael, had to be called down to the docks from another manual job he was working in the offseason. His connection with "screen tourism" was as an additional source of customers, but looked at in this way, especially with the expansive definition of "media" and "creative" industries we have seen in the last two chapters, his exclusion from these frameworks is worth questioning.

Vicki Mayer critiques the class and geographical gatekeeping of creativity by dominant "creative industry" models, looking at media workers excluded from dominant narratives of creative and cognitive work. She demonstrates that television assembly line workers in Manaus, Brazil, for example, are as much creative workers as above-the-line television "producers" in the United States (an ironic semantic framing not lost on Mayer or her interlocutors) (2011, 31–65). But studies in creative industries have been predominantly urban in character, spatially biased and failing to account for the rural and suburban character of many spaces of creative production. Against this prejudice, the Ireland 2040 framework articulates increasing rural focus to creative industries: "Rural areas have significant potential in these sectors, and as digital links and opportunities for remote working and new enterprises continue to grow, employment is likely to increase in areas such as agri-tech, ICT, multi-media and creative sectors, tourism, and an added value bio-economy and circular economy" (Government of Ireland 2018b, 75). Policy reports like those of Creative West (White 2010), released in the depths of the financial crisis, inform state decisions about policy governance of rural regions and demonstrate the stakes of economically stimulating Ireland's expansive rural areas in the face of global financial turbulence and other industrial changes. As the report states, "Given declines in rural employment in agriculture, traditional manufacturing and construction, the creation of alternative employment options in rural areas is fundamental to their continuing viability" (82). But within this rural creative ecosystem imagined by planners and policymakers, the enterprise of local business owners, operators, technicians, laborers, and artisans should be seen within the wider remit of creativity, which typically includes more traditional media, high-tech, and "highly skilled" work. The language employed polices distinctions between more traditionally creative (and profitable) cultural work, foreclosing potential aid to these so-called ripple and spillover local service and blue-collar industries.

This is not to say that more localized proposals and enterprises to make media industries more permanent, robust, and beneficial to local talent and operators in the southwest region are not afoot in Kerry. A film studios project, endorsed by the county council, sought planning permission for an industrial park outside of Tralee, but I was told by its backer that "the site is parked at the moment" (as of winter 2019). Endorsement by government authorities did not necessarily translate into the required investment. Screen Kerry, a local film commission funded by Kerry County Council and the Kerry Education and Training Board, promoted skills training and location management in the region. The 5 percent regional uplift to the S481 was seen as a huge boon for a countywide project of growing the film industry among those I spoke with, but the novelty of the scheme at the time and the realities of infrastructural access to the region meant that few projects had yet made use of it, and its yearly scaling back obviously affected the circulation of "481 spend" through rural Kerry through the scheme's duration and its aftermath.[20]

An instructive model of a rural creative industries experiment is the Dingle Hub, an enterprise and coworking space that boasted a diversity of "creative" projects when I visited in 2019 (figure 3.4). A creative industry outpost nestled against the small fishing and tourism port in Dingle's town center, the building, shared with the Dingle ÚnG offices and that visibly resembled a community center more than a tech and media start-up hub, was hosting workshops and advertising itself to the attendees of the Dingle International Film Festival (DIFF) and Animation Dingle being held on the March weekend I arrived (figure 3.5). At that time, it was shared by workers from journalists to composers to biologists and even a software developer working in just-in-time software logistics, among a hodgepodge of other industries. It boasted the "fastest broadband in Ireland" due to eir's experimental presence in the area. They were also advocating to expand into the disused Dingle Hospital and Workhouse, proposing that current administrators, the Health Services Executive, transfer the land and facilities to ÚnG to host a new "creativity and innovation hub" housing media workspaces, arts studios, and other community and tourist facilities.

But even without such incentivized creative industrial development, Dingle has a long history as a cultural hub, both from local talent and outside transplants settling there, tied to the region's spectacular landscape and Gaeltacht culture. Film and arts festivals regularly enliven the town. However, these short-term events have more to do with the service and tourism industries than what are typically considered within creative industries

3.4 Entryway to the Dingle Hub office park, which it shares with ÚnG, in winter 2019. Photo by Patrick Brodie.

3.5 Brochure handed out to attendees of Animation Dingle and DIFF in winter 2019. Photo by Patrick Brodie.

policies, which require an ongoing series of such events to sustain such an economy. Any harmonious picture of an artistic community inviting new visitors to a burgeoning tech and media ecosystem masks these significant underlying problems. The *Star Wars* tour driver informed me of the difficulties in settling there for both new arrivals and local community members. There was a serious housing shortage in Dingle, and even tourists often found it difficult to secure accommodation in the small town during festival season.[21] Acquiring property across the peninsula had been made more difficult by the boom-era landgrab that saw Irish city dwellers and European holiday makers buying up property for new builds and vacation homes. A planning loophole was closed, which made it more difficult for this to happen, in response to a quite unregulated period of building. In addition, the fact that nearly all Irish coastline is regulated as "green space" and few new builds are allowed (although old cottages can still be modified in their original footprint), and the small size and thus limited availability of housing in towns and villages, meant that the gamble to build a creative hub in Dingle was much more complicated than getting media workers and holiday makers to stay for good. It would require a fundamental change to the area's housing situation and spatial planning, which was facing a perfect storm of tight regulation due to heritage and local protection and a countrywide housing market favoring short-term rentals and vulture property investment. Airbnbs dotted the peninsula, but permanent housing was scarce and expensive. In this small town of roughly 2,000 residents, as far west as you can get in Ireland, the spatial economy of the "creative city" described in chapter 2 had unexpected resonance.

By typical creative industries logic, like Troy Studios, Dingle Hub needed to attract a core pool of creative workers to sustain its enterprise goals. While the region has its share of artists and craftspeople, film and media industries needed to be imported, along with the requisite workers to sustain them. As Pauline White reports in her policy paper on rural creative industries in the west of Ireland, "the most critical questions are what attracts creative talent to rural areas and how the creative sector can be supported and developed to become a more significant component of the rural economy" (2010, 82). By these models, the Dingle Hub would ultimately need to attract new "talent" and "skills" to the region in order to stimulate the ecosystem, and one of the primary draws was the idyllic location. The creativity consultant and project manager at Dingle Hub at the time had moved to the peninsula for exactly this reason. While several of the folks in Dingle Hub were "locals," even more were blow-ins and transplants.

Nonetheless, global productions (and tourism-inviting events like Animation Dingle and DIFF) were disruptive to the very "rural character" that attracted people there in the first place, let alone to the locals who make a living in the service industry, tourism, or otherwise in the region. The creativity consultant and project manager lived in Ballyferriter, near Ceann Sibéal where *The Last Jedi* filmed its scenes on the planet Ahch-to. She said that the filming unsettled the usually quiet area, with helicopters and lights igniting the night for several days. While the large production was a boon for local business in the short term (and the wrap party at Foxy John's in Dingle celebrated everyone involved in the wider community), the monthlong set construction was punctuated by a comparatively short shoot (around a week). During such periods of heavy traffic and use, existing local infrastructure tends to break down. While I was on a national bus from Tralee to Dingle in 2019, the driver had to ask locals to disembark and find local transportation in order to accommodate the DIFF and Animation Dingle passengers (including myself). The *Star Wars* tour driver similarly warned us to watch out for American drivers rocketing around the turns when we would make roadside stops around the peninsula, as the narrow ring road's inclusion on the Wild Atlantic Way drive tourism route failed to mention an informal local custom to drive in one direction depending on the time of day.

This raises important questions as to strategies for cultivating local talent and infrastructure by first attracting outside investment—whether from FDI, tourists, or permanent transplants—with the "local flavor" that nourishes the place brand. The limited infrastructure and "wild" landscapes of the region cannot be expected to accommodate a sustainable abundance of large-scale productions, let alone a viable workforce at scale, without significant other investments in infrastructure and public services, which, as chapter 2 explains, "creative industries" policy leaves to the market. While *Star Wars* made use of Kerry's spectacular landscape as a stand-in for otherworldly settings, global production flows follow place-agnostic logics of circulation and movement that nonetheless must manage conditions on the ground.[22] This is typically achieved by flattening contingency through short-term management and by extracting value from local labor, culture, and landscapes. However, as my interlocutors made clear, this is not always a one-way process: The tourism economy in west Kerry had created remarkable community-led and uncanny encounters between Irish-language culture and the global brand of *Star Wars*.[23] Still, these arrangements were precariously placed within the aftermath of runaway production, founded on the exploitation of localized cultural labor, and unsupported by public investment. One Airbnb host, involved in the annual May the Fourth Festival in Kerry, made clear that despite seeking funding from tourism bodies, the folks sustaining *Star Wars* screen tourism were largely doing so on their own.

Creative industries policy tends to see such precarious workers and those not represented above the line as supported by spillover effects, rather than paying attention to what they actually do or need. This is because, under prevailing logics of global capitalism, those not accounted for are seen as freely available labor pools, resources at the ready for exploitation. As at the global scale of the race to the bottom and transnational competition for FDI, conditions of precarity are most strongly felt by workers who are left with no choice but to compete with one another in the absence of strong social infrastructures of care and collectivity. A better political economy accounting for policy's meaningful encounters with labor requires engaging deeply with the precarity structured into the creative industries and the media infrastructural formations that it gives rise to. If the Dingle Peninsula and west Kerry in general demonstrate the eventful interactions between global modes of production and deeply emplaced cultures and landscapes, we can perhaps think about the ways in which culture, labor, investment, and the natural environment are differentially treated and infrastructurally managed as "resources" within the policy logic of nation-states and the operations of capitalism.

Naturalization and Precarious Media Labor

What film production and its externalities in places like Kerry can tell us is that what the global race to the bottom looks like, then, is not only place-averse global productions and enterprises simply using a space for its placeless and modular properties, as in the case of film studios' role in runaway and free zone production in the Global South (see Curtin 2016, 677; Dickinson 2024). It appears as the high-water mark of capital investment left behind, the residual industries and economies that form, and the resulting labor implications and exploitations that arise in the vacuum of investment, as much as it does the ongoing ebb and flow of global production and its more eventful forms of exploitation. As the previous section exposes, the spatial logics of film and media policy demonstrate a management of labor pools similar to that which captures value in the creative city. Across policy literature, whether that of media production or the creative industries more broadly, workers are treated differentially depending on their role and its perceived value in the larger economy. For example, creative industries policy privileges the needs of "creative" professionals within an often unclear range of industries (from the arts to finance); film and media policy sees ripple effects and spillover industries like screen tourism as secondary to the initial facilitation of production, leaving its actual governance to tourism bodies; and media production more generally treats "above-the-line" talent (producers, actors, directors, screenwriters) as more important (and better paid) than "below-the-line" workers (gaffers, grips, set decorators, costume designers, lighting technicians, drivers, caterers), reproducing an on- and off-set class hierarchy that informs how workers are treated within the production and policy equations of national and transnational organizations.

Part of this comes from the neoliberal classification of workers under metrics of "cultural tests," "eligible spend," "human capital," and, most pertinently, "talent" and "skills." These are not external equations to transnational productions, but rather increasingly treated as resources to be *managed* within production via efficient media policy and business. How much Irish spend can you maneuver, and how do you game the measurements to ensure you wring the most value out of the partnership?[24] In these logics, human labor is accounted for as *spend* in the same terms as building use and technologies, and the metrics of jobs meant to account for public processes and funds that disappear into the pockets of investors through financial manipulation. The front page of SI's website was still in 2023 organized around three pillars, featuring stills from successful Irish films: Talent, Creativity, Enterprise (Fís Éireann/Screen Ireland 2020). These are familiar categories across creative industries discourse. Talent: the labor pool is educated, and biopolitically primed to work. Creativity: immaterial labor, ready to be harvested and exported. Enterprise: foregrounding the business of cultural work, or to quote Michael Curtin and Kevin Sanson, "organized according to industrial principles that economize at every step in a sprawling creative process, constantly seeking efficiencies and accelerating workflows" (2017, 1).

But in the film and media sector specifically, knowledge production and dissemination are articulated in terms of "skills" and "talent" as a way to manage labor bases. Screen Skills Ireland (SSI), a training body established independently in 1995 but now attached to SI, advertised itself in the late 2010s as an agency designed to "grow the sector" and provide the conditions for it to "thrive," attaching both the active and unenfolded workers of the country to a project of economic growth by providing them learning opportunities for a fee. ssi workshops included anything from producing, to writing, to innovation, to "ambition" workshops, demonstrating the affective and cognitive elements of labor skills thought to be required by such training organizations. They pitched these toward global and transferable skills for the "sector" broadly conceived (Screen Skills Ireland n.d.). By advertising national skill sets to foreign investors, while simultaneously training the workforce to accommodate the industry, si treated its native workforce like a reserve army of labor at the ready for global production, actively building "skills" to be used in the sector. However, occluding the distinctions within the workforce, the website painted workers, managers, and executives based in Ireland with the same brush as broadly defined "stakeholders," even though, in their own words, they needed to appease market forces with workshops that responded to "skills shortages" determined by global producers (however much these might also be identified by workers needing to round out their technical capabilities to gain in this economic system). ssi workshops were often held online or at various locations across the country, representing both the flexibility and regional unevenness of the media workforce and the infrastructures available for professionalization (and also showing which places had been deemed strategic by industry advocates and policy agencies). For example, they were frequently held at private film infrastructures like Troy, designed, like the Dingle Hub, to build up local infrastructure and labor resources by facilitating the flow of transnational capital and drawing in workers in tandem.

This preceding paragraph should not be read as a critique of those who promote and use such skills-training mechanisms. To ssi's great credit, they would offer workshops based on requests from industry professionals, and such skills development agencies offer an indispensable architecture for supporting film and media artistry and professionalization. However, just as I do not advocate reducing, but rather increasing, direct public funding for film and media more broadly, such training mechanisms must be led positively by the democratic appeals of workers and artists to cultivate meaningful livelihoods and cultural expression, rather than responding to "skills shortages" defined by global production markets. Kay Dickinson positions such skills training organizations within wider transformations in media and education to accommodate global supply chains in given policy environments with cheap, trained, and casualized workers (in her study, the UK) (2024). As more and more skills are required to perform even entry-level roles in these industries, these continuing education and training courses are part of wider social and industrial shifts toward on-going professionalization that devalue increasingly expensive degrees and certification programs offered by universities while accommodating the continuous flexibility and upgrading of staff required by a globalized media environment (see Bousquet 2008; Harney and Moten 2013; Levidow 2002). In short, skills development agencies like SSI provide an essential service for workers. However, by emphasizing "skills" to facilitate global media operations, they reflect troubling elements of the biopolitical management of labor by the state and the subjection of workers to place-agnostic requirements wholly determined by capital.

By leaving workers in a place exposed to the whims of global production marketplaces, especially in terms of tax breaks, on-the-ground labor and management are always going to be subject to exploitation from afar and, in the process, precariously reliant on legislators remaining friendly to international production. For Curtin, "this geographic mobility and flexibility that has come to characterize big-budget feature production is crucially reliant on location staff that serve as an interface between the global apparatus and local resources. They make the system viable, and quite crucially, profitable. Without them tax incentives would not be as valuable to Hollywood producers or to the state governments that offer them" (2016, 678). In short, as Curtin states, it is not hard to imagine what would happen in these locations if "state legislators suddenly turned off the tap" (680). But as he continues, so-called "talent" and "creativity" are increasingly sources of value for extraction by transnational media companies (681), positions echoed by the definition of the "audiovisual industries" in the 2011 *Creative Capital* report, commissioned by state and industry stakeholders at the height of the crisis, as "a dynamic mix of talent (human capital), creativity (ideas and innovation) and enterprise (adding and extracting value)" (Department for Arts, Culture, and the Gaeltacht 2011, 27). Cultural labor forces and infrastructures are treated as tools for these economistic logics. Media policies continue to "dole out tax breaks and strengthen their institutional infrastructures" (Curtin 2016, 681) to attract investment to cultivate (and enable the extraction of value from) these resources. However, policymakers extolling these logics fundamentally miss the central biopolitical dimensions of labor pools managed within their territorial governance.

Curtin's association of funding and talent with another infrastructural resource—water—interestingly echoes the language of tourism operators in Kerry. The *Star Wars* tour operator, when discussing the slow months of the winter tourism season (when I visited) in contrast to the busy summer, told me that it was like "turning on the tap" once the seasons changed. A similar image was conjured on a slow day in Portmagee by a boat captain who took visitors to Skellig Michael and painted me a picture of the tiny main strip in the town (a shop and two pubs) laden with tourists in line to sign up for the limited trips out to the island. Even the idea of ripple or spillover effects coming from top-down film and media policy initiatives, like the central metaphor of the "wild tides" of global capital that informs this book, refers to flows of water generating surface effects far from their source of contact—or even literally brimming with abundance for the global production while overflowing droplets unevenly nourish those around the edges.

Natural associations abound within the policy logics required to naturalize the place of FDI within given economic climates (as will be unpacked in chapter 4). The circular logic and association of resource health and distribution that characterizes the FDI-driven logics of Irish media production, where foreign capital, the landscape, infrastructure, and the labor force are seen to partake within a mutually beneficial ecosystem, fundamentally cannot account for the problem of global infrastructures that maintain spatially uneven and distributed exploitation. In participating in the naturalizing language and metaphors of this distributed supply chain of global media, Irish film and media policy positions the people and the landscape within the country as mere resources available for extraction and exploitation by transnational forces. Imagining the flows of capital as somehow akin to water trickling down or being piped in from above, sustaining a cultivated ecosystem, means that at any moment, entire industries and regions can be left arid if the tap stops running for any reason—whether if turned off by policymakers, the flow redirected, or the supply runs out.

As the setup of this chapter emphasizes, the redirection of capital flows—and the contracting of care to the private sector through film and media policy to attract multinational investment—hands control of the industry, and thus the labor market, to precarious and frequently exploitative flows of investment. As Mayer reminds us in her creative analysis of a regional tax break in Louisiana (2017, ix–xi), capital is not natural. Owners of capital work to exploit infrastructural, governmental, and labor systems to redirect this value further and deeper into their own pools of accumulation, while leaving the people and places that generate their wealth depleted, high and dry. While Mayer plays on the spectral quality of a roving tax incentive as a cynical joke, the effects of these economistic treatments of culture, space, and labor are deadly serious. With the equations of policymakers and legislators veering further away from questions of cultural identity as media and film cease to be purveyors of public good, the biopolitical metrics of "talent" within labor pools restructure how film and media are governed by positioning meritocratic lines around these abstract categories. As Marxists of the autonomia tradition have decried, this biopolitical restructuring of labor via "passion" and cognition betray technocratic mechanisms of intensified labor exploitation in the creative industries across sophisticated technological networks and systems (Berardi 2009; Lazzarato 1996; Terranova 2000). What Curtin recognizes as the "confusion" of policy officials between "cultural concerns and economic and political objectives" (2016, 683) is rather a decision based on a logic of austerity that positions people as capital—contributors to a machine of transnational extraction—within an intensified environment of entrepreneurialism and individual realization, at the same time that the official state line is that these industries are essential for the livelihood and well-being of entire pools of workers. In these apparently competing objectives, competition over apparently scant resources means to atomize rather than unify workers and communities, contributing to the ongoing tensions that maintain the status quo.

However, despite the clear creep of biopolitical governance into film and media policy via "creative industries" discourses (which, as we saw in chapter 2, are deeply spatialized), Curtin's policy recommendation thus moves toward treating local labor pools in terms of "resource management," as a recognition of multiplicity and difference within a given territorial labor pool. He argues that policymakers "might on the one hand blend public and private resources, while on the other help to sustain micro communities or oppositional constituencies that have absolutely no commercial value. Media should furthermore be characterized—like any healthy ecosystem—by tension and antagonism, as well as interdependence and symbiosis" (2016, 683). While this is compelling, it seems to me, to borrow a phrase used by Jennifer Holt and Patrick Vonderau, a kind of "regulatory hangover" (2015), demonstrating a faith in nation-states (and governance in general) to assert cultural autonomy from the intrusive economic forces of globalization and the intensified labor exploitation of neoliberal economics. Beyond a perhaps optimistic sense that a different film culture might emerge through such a regulatory infrastructure, this return to an idea that regulation *can* happen in a normal, healthy global ecosystem is one that has been continuously foreclosed by the coalition of capital and the state under transnational organizations of power.

Such a "public resource" logic surrounding labor might actually mesh with the existing rural film and media sector in Connemara, a Gaeltacht region west of Galway (see Power and Collins 2021). Radical Irish-language filmmaker Bob Quinn famously located his production company, Cinegael, in Carraroe in 1973, and the region has maintained semicontinuous film activity since. Many interlocutors assured me that this was the true center of media production in the Galway region. Here, the sector has been well established, with operational film studios (Telegael), along with a thriving arts community centered around its active Gaeltacht and the cultural activity and programs surrounding it. Since 1996, the area has been home to the main offices of TG4, the country's national Irish-language broadcaster, and the small town of Spiddal hosts film studios Telegael and the permanent set of long-running Irish-language soap opera *Ros na Rún* (1996–present). While foreign and coproductions were frequent, with "481 spend" becoming more prominent during my research, the industry in Connemara remains largely maintained by its services to the national media industry (from TV, to commercials, to other content).

Ros na Rún's production also formed the basis of a robust training infrastructure for talent in the region and provided long-term permanent employment for a variety of workers and technicians, if you were lucky enough to take over the spot of anyone retiring or moving on to a more lucrative gig elsewhere. An internship program to apprentice with camera operators, lighting and sound technicians, and other technical roles built up skills for local workers, but would not usually result in continuous employment afterward. Similarly, Roger Corman's establishment of a film studio in Spiddal in the mid-1990s, attracted by tax incentives to make his cheap B movies, contributed to the training of much of the local labor force.[25] Nonetheless, coupled with the activity and funding of ÚnG and other Irish-language schemes, the Galway region and Connemara offer a fascinating foil to FDI-driven fortunes of other regions in the country. While not a perfect or even all-encompassing media system for the region, and despite the pervasiveness of freelancing and contract work present there as elsewhere, there appeared to be the groundwork for more sustainable programs in place to ensure that cultural and media production prevailed, even if it often had more to do with the protected status of the Irish language than an across-the-board cultivation of the arts and cultural activity.

While examples like Connemara and Curtin's argument are compelling and seductive from a regulatory standpoint, such formations do not exist without tension from capital, and demonstrate the ways in which potential complacency might allow deeper imbrication of culture and capital. Curtin's approach assumes a circulatory capitalism by which the market generates its own "first nature," to use Hegelian terminology analyzed by Neil Smith (2008), describing a process by which the social conditions of global capitalism, and the market along with them, come to appear as naturally occurring phenomena—similar to the immanent logic of circulation-based capitalism outlined in chapter 1. While it is undoubtedly true that capital naturalizes operations through a variety of mechanisms, treating labor markets, laborers, and local contingencies as "resources" in a global ecosystem of production (along lines of circulation) seems a troubling naturalization of global productive forces that are at their very center based on spatial and labor exploitation.

The move to such natural associations is understandable under the logic of "sustainable" forms of development, especially in terms of the (comparatively) modest goal of damage control, of preventing short-term exploitation and mitigating the exhaustion of cultural supply lines and labor markets. However, in an environment of austerity, during which Ireland undertook *forcible* privatizations across its public sectors and *obligatory* cuts to public programs and cultural budgets, which effectively required privatization and a deeply permissive attitude toward FDI (including natural resources such as water and forests), the idea of a return to regulatory strength—especially within resource logics—via a segmental shift in state logic seems far beyond the realm of possibility. Even in Connemara, film facilities were (and remain) increasingly in use as channels for 481 spend, demonstrating the geographical spread of extractive FDI even to areas with more robust public infrastructures. SI rents fragile cultural and environmental heritage to Disney; environmental resource agencies like Coillte sell forests to Apple (see chapter 4). Any conclusion to treat local labor markets—and the ecosystem of global labor in general—as a "resource" tasks policymakers with sustainability in a way that specifically enables an explicitly extractive economic mindset to govern workers in the film sector, in ways that extend even beyond "creative industries" due to the added precarity of culture as a previously "public good."

What, then, does this mean for workers and those concerned with the industry more broadly? To avoid overgeneralizing the Irish case, what we should note is that film and media policy that is designed as damage control for the continuous restructuring of the global economy must always take exploitative production for granted as a necessary fact of governance. Policy oriented toward the FDI-driven logics of global media production, tourism, and to a lesser extent coproductions always derives from a place that situates a country competitively within a global race to the bottom that puts worker stability, culture, and public good at the margins of broader economic concerns. In Ireland, this has led to an environment in which workers barely see themselves represented within policy in ways *other than* resources for producing value, or as external, tertiary beneficiaries of top-down investment. This does not necessarily feed into how they see themselves in the industry—far from it, as they tend to view policy concerns as peripheral to how they *actually* make a living and engage in cultural activity and enterprise. However, the approach to the phenomena presented in this chapter reveals something troubling about Irish governance, using Curtin's idea of labor as a "resource" and confronting how these policy logics represent workers (and spaces) within them as upgradable and manageable bundles of talent and skills for use by multinational investment. It demonstrates an ongoing extractive logic, driven by a long history of FDI and the treatment of land by the Irish state, that cannot but utilize available "resources" for exported profits.

Conclusion: Incompatible Alignments

Speaking to Aosdána, an elite national assembly of artists, the Uachtarán (president) and sociologist-in-chief Michael D. Higgins's 2015 remarks framed the place of culture within the logics of economic development. He stressed that creativity and culture were about the right for everyone to participate in a society: "They are a social good which, if left to the vagaries of the marketplace, will either fail to survive or become so compromised and distorted that the public good will not be served." He continued, "As a society we must come to recognise that institutional provision for the arts is as important to our infrastructure as roads, hospitals and schools," embedding cultural production within national infrastructure. Finally, he noted, "While culture has its instrumental uses, including employment creation, we also have to recognise the limits to which this can be applied without endangering it. The arts are increasingly recognised as useful in the sphere of economic development at home and as burnishing our prestige as a country abroad. But the arts are not merely a tool of tourism or commerce. Yes there is of course an economic case for arts expenditure but that case is not, and cannot be allowed to become, the only reason for such expenditure" (2015). While these remarks are admirable, they ultimately represent the limits of critique when maintaining ideals of rights, public, representation, and citizenship within a liberal, market-driven economy and society. Admitting the instrumental uses of culture demonstrates the fact that these logics have already crept so far into public policy that there is no return without radical overhaul, outside of the remit of civil society organizations as they are currently constructed. Just as capital caused the crisis, it profits from the response. Creativity and culture—through the creative industries—represent the logical point at which a crisis of finance and feeling redirects the excess capital of "public" good back to the private sector, with a remarkable sleight of hand.

However, despite the macroeconomic and geopolitical influence of transnational and global media economies across Irish space, this landscape is still not *smooth* or *organized*, but lumpy, ad hoc, uncoordinated, a heterogeneous assortment of producers, service providers, unions, public and private agencies, workers, and studios, especially at the bottom end of (or excluded from) industry accounting sheets. Offshoring and supply chains are processes of circulation as much as production (see Urry 2014) and exist in direct relation to transnational finance and the state's privileging of this fictitious capital. Transnational interdependence operates similarly. As Mezzadra and Neilson articulate, the intertwined fields of "extraction, logistics, and finance" within the current global structures of power mean that their inseparability should inform the study of any one individually (2019). But while "extraction" has taken on conceptual provenance in recent political economies and theories, Mezzadra and Neilson are careful to warn us not to forget about the centrality of "exploitation" even within decentralized networks of extraction, and emphasize the gendered and racialized processes of violence at the heart of these assemblages of power and domination that occur at the nexus of the state and capital (2019, 201–8). In doing so, they articulate how forgetting "exploitation" that occurs across extractive processes and frontiers (from logistical networks to financial calculations) risks foreclosing radical alternatives within which social cooperation is *not* enmeshed within systems of capital and collective action *is* possible.[26] In this environment, we need new words and new approaches to film and media economies that do not abstract processes as somehow transcendent or top down. We must certainly understand how FDI parachutes in from above; but we need to also understand how transnational capital's coordination on the ground can reveal antagonisms toward a more equitable and just media system, opening avenues for resistance and alliances *against* the commonsensically unjust alignments of state and capital.

Film and media policy and its transnational economic arrangements clearly have a material and spatial politics. However, it is a politics where decisions occur in a transnationally oriented, circulatory sphere within which precarious workers have little or no democratic input. What became clear in my travels across Ireland to speak with media workers and attend media events was that the prospects of "scaling up" to even the national scale for local workers (let alone a transnational or global one), which was the entire policy justification and ground game for the Irish film industry, were extremely unlikely, the paths to even modest "success" treacherous and unclear. Film and media policy does not in fact generate the kinds of localized effects that are put forth by wider rhetorics of jobs, ripple effects, trickle-downs, and spinoffs. It is not a unilaterally successful or robustly sustainable way of bringing production in. It does not even, at minimum, ensure the employment of a localized labor force, an outcome promoted by cultural funding agencies in terms of justifying policy to sustain the industry (whether Irish bodies or European ones). Of the roughly twenty government agencies, arts and media institutions, production or postproduction companies, and media workers that I visited and discuss in this chapter from 2017 to 2019, the vast majority used or worked primarily with and through short-term contract labor.

So, if the private-sector bodies that arrive to make larger-scale projects are meant to provide social good and sustainable futures via consistent employment, technical infrastructure, and skills training, this falls apart once you start to ask even basic questions. Why did Disney say that Ireland needs more skills training and better infrastructure if their investment was itself meant to grow these material needs? On a level of basic survival for workers and artists, what were these project-based employment opportunities going to provide in between big projects? Needs were always determined by market forces acting in an ecosystem that is dictated by corporate executives and policy officials (with limited engagement in this low-level work). The naturalization of these market forces as the lifeblood of sustainable media economies represents a dangerous development toward corporate sovereignty over what may have once been treated as public goods. The effects on labor are profound and explicit, and require critical attention beyond media studies—this is about a reorganization of cultural, "creative" economies that can be observed across sectors and need radical shifts in policy orthodoxies to correct.

Digital media and content companies could not survive without the robust distribution apparatuses offered by transnational technology and logistics companies like Amazon, Apple, Facebook, Google, and Netflix. Most of these companies are today content producers in their own rights, integrating the media supply chain within their expansive operations. These processes have been highly transformative, in terms of both production and circulation. As capitalism's center of gravity has shifted from production to circulation, Ireland's policies of supply chain facilitation for FDI have made it an ideal place for companies to locate these operations, especially in the speculative, enthusiastically investment-hungry early stages of the financial crisis and its aftermath. Building on insights about the naturalization of these logics within media policy in this chapter, in the final chapter I will analyze the instrumentation of this pervasive business and natural "climate" in Ireland by the "data center industry" and its media infrastructural apparatuses.

4——Storm Clouds

Technology Industries and the Climate of Crisis

One morning in the summer of 2018, I was picked up at the Athenry passenger rail station by Seán,[1] an organizing member of the Athenry for Apple community advocacy group. We had arranged a site visit to Derrydonnell Woods, where Apple, after several years of dealing with environmental objections through the complex Irish planning system, had recently withdrawn a proposal to build an €850 million data center campus, which would have constituted the largest private influx of capital in the region's history (O'Donoghue 2017). The location in east Galway was particularly significant from a development perspective, as an oft-neglected region of the rural west, more closely aligned with the flat, boggy Midlands than the mountainous landscapes and Gaeltachts of the far western coastal regions along the state-branded Wild Atlantic Way tourism route. Seán and I had only met via Facebook before this point, so we made small talk during the ten-minute drive out to the woodland site on the outskirts of town. He told me his family had been in the area for generations. As we drove past the Gaelic Athletic Association (GAA) pitch, near the edge of town, he remarked that his grandfather had been among the small contingent of local rebels to gather there during the 1916

Easter Rising. This historical detail, unrelated to our journey that day, was typical of the living character of history in Ireland, where the past is a readily accessed and remembered presence in the landscape, always resurfacing and coexisting with ongoing present circumstances and potential futures.

When we arrived at Derrydonnell Woods, we parked at the end of a small unpaved access road, marked as private property, with a standard advisory that the "occupier," the holder of the land, had no "duty of care" toward visitors. However, there was a well-trodden woodland trail, and we encountered one or two other small groups of walkers as he showed me the site. It was a hot and sunny summer day, especially by the standards of the west of Ireland, but the woods remained cool. This, Seán informed me, was one of the primary draws of a woodland site for a data center, where the air temperature would be several degrees cooler than even the surrounding landscape, blanketed much of the year by rain and wind. He showed me a variety of markers along the path, including one that monitored the shallow water table. A data center would apparently have had little risk of affecting the groundwater, as opposed to a trash incinerator, which had once been planned for the site, something that could have released and circulated toxic particles through the area underground. Seán and his family lived adjacent to the land, and water integrity had been very important for him when considering his support for Apple's proposal. We eventually came to an artificial clearing in the woods—most of the trees appeared to have been cut down, and those still standing were dead or decrepit. Coillte, the semi-state forestry company that had facilitated the land handover to Apple, had once overseen industrial logging of the woods, leading to swaths of the 500-acre forest being cut down for lumber (figure 4.1). Woods are not terribly widespread in contemporary Ireland outside of conifer plantations, so those that remain or have been replanted are quite cherished.[2] Seán told me that this logging had already done damage to the ecological and aesthetic integrity of the area, and that Apple, which had planned to build public walking trails around their private site through the woods, would have helped restore the area to its former grandeur.

The suspended future of the site, still tied up in decisions being made in Apple boardrooms in Cork, at the closest, or more likely in California, was felt most palpably when we returned to Athenry town center and debriefed over coffee later that morning. Speaking with the barman at the local pub we landed in, who had also been involved in Athenry for Apple, among others that passed through briefly or sat sipping their Guinness and other drinks across the bar, the hope that had once flooded the town in anticipation of

4.1 An apparently logged portion of the Derrydonnell Woods site, summer 2018. Photo by Patrick Brodie.

the project had manifested in defeat and bitterness. Just a few years after 2,000 residents (for reference, Athenry's population is roughly 4,000) had taken to the streets to demonstrate support for the project, the town felt drained. The community had come to distrust especially the objectors to the site, two of whom, the principal objectors Sinéad Fitzpatrick and Allan Daly, lived locally. Daly, however, was not a "local"—an engineer, he came from the United States originally, but was married to a Galwegian involved in the city's arts community. After Seán left, I stuck around and chatted with the barman, who, with a meaningful look, suggested I dig deeper into Daly's affiliations: "Who's funding Daly?" He said that no one could discern the American's motivations. When I contested and said that Daly's concerns seemed genuinely environmental, he was doubtful.[3] At other times, however, he was more hopeful about the ability to overcome the division and abjection felt by those in the town. "We'll battle on, hopefully the wheel will turn." He pointed me to various storefronts in the town where he said that residents involved with Athenry for Apple would be happy to speak with me. But on the contrary, most whom I encountered were sick of talking about it. An "Apple hangover," as one person said, had taken over. Apple, though, was spared most of the bitterness, which was directed primarily toward objectors and a woefully inefficient state.

The failure of Apple's project in Athenry was widely associated in the press with an imaginary of the west of Ireland as a place of messy cultural politics and difficult planning conditions (see Paul 2016). The town's ambiguous location between the scenic west of Ireland and the Midlands, existing between the "place-branded" (Greenberg 2004) jurisdictions of Fáilte Ireland's Wild Atlantic Way (see chapter 3) and newer Ireland's Hidden Heartlands campaign in the Midlands, among other gripes with the state's lack of meaningful focus on rural Ireland, had left the area largely deprived of the creative-industry and tech-led "recovery" happening elsewhere across the country. Recalling the histories of uneven (post)developmentalism during the Celtic Tiger and post–financial crisis aftermath described in chapter 1, Athenry in 2018 was one of many towns in the rural west still reeling from recession. The final Supreme Court ruling on Apple's proposal, concluded in spring 2019, long after the company had withdrawn their proposal, was even, exceptionally, held for public attendance at Aula Maxima on the National University of Ireland's Galway campus, a spectacle of government offered by the state to frustrated residents in the west who felt largely forgotten and abandoned. As journalist Mark Paul quipped in the *Irish Times*, Apple had been "stuck in mud in the fields of Athenry"

(Paul 2016), a dual reference to the boggy imaginaries of Ireland's rural regions and the famous rebel tune "Fields of Athenry." This tongue-in-cheek headline encapsulates, intentionally or not, many of the complexities of culture, history, capital, politics, and the environment being entangled at the site. Tying the data center to a complicated tangle of cultural and environmental histories in Athenry, to the rest of the country, the aspiration of global connectivity was lost in the peaty mud of the western Irish bogs, the suffocating thickness of life and territory in the west too dense to crawl out of once immersed.[4]

However, what is perhaps lost in this narrative of failure is that the general populace of Athenry, with many supporters across the region more broadly, was *actively and enthusiastically in support* of Apple's project. It was blocked by a small minority of objectors, who contested the project's environmental toll, which, across the appeals process, included disruptions to the local ecosystem, noise and light pollution, and the astronomical energy costs disregarded in planning approvals and environmental impact assessments (EIAs) carried out by An Bord Pleanála.[5]

Athenry offers a reference point for this chapter through which to understand the entangled media infrastrutural politics of post–financial crisis Ireland through the lens of multinational tech corporations, their logics of value, and their politics and operations on the ground. To borrow a phrase from Jennifer Holt and Patrick Vonderau, Athenry provides a fecund example of "where [and how] 'the cloud' touches the ground" (2015, 75). These "peripheral" places across rural Ireland, the subject of much of chapter 3, have been imagined as frontiers for economic development for decades, as chapter 1 contends, despite the messy and uneven dynamics that tend to slow, stall, or prevent these projects. The cultural and environmental politics conjured by Apple, the state, the media, and the residents of Athenry, as other examples from each of the prior three chapters have suggested, is not only serendipitous. These politics represent the entangled logics of value that see Irish space and culture as resources for multinational corporate exploitation and extraction, in ways that have pervaded the public sphere.

Financialization and austerity have led to a widespread (if sometimes reluctant) acceptance of certain economic realities and modes of production, and these are tied to much longer histories. Far from encountering "disempowered" populations, however, these projects continue to be thwarted or enriched by cultural and community politics on the ground and through the organizational and spatial structures of data centers. Anna Tsing tells us that "supply chains don't merely use preexisting diversity;

they also revitalize and create niche segregation through advising economic performance. Understanding supply chain diversity, I argue, requires attention to niche-segregating performances; such attention, in turn, should advise our analysis of the *global* in global capitalism" (2009). Attending to the vital diversity of on-the-ground politics factors in how the "global" is coproduced and enriched by localized experiences of material circulation. In Ireland's austerity-driven recovery era, these experiences provided experimental conditions by which data centers emerged as mechanized infrastructures for economic development led by multinational tech, centralizing arguments surrounding foreign direct investment (FDI), media infrastructure, creative industries, and environmental transition within these corporate technologies.

This chapter argues that Ireland's post–financial crisis media infrastructural environment contributed not only to a global economy of digital media industries, but more specifically to an emerging global climatological politics through which value can be extracted by transnational corporations in times of crisis and transition. Climates are not only meteorological. They are socially constructed, embedded within histories of capitalism and colonialism, and thus in many ways subject to human intervention and manufacturing (Chakrabarty 2009; Furuhata 2022; Hulme 2017). Beyond gaseous and particulate atmospheres, climates consist of business interests, tax structures, policies and planning, infrastructural arrangements, civil society, and cultural politics, which are entangled with local environments, ecologies, air, and geologies in profound and often unexpected ways. Climates are effectively the array of naturalized conditions through which atmospheric change is registered. But they are also made and intervened in by states and capital. Taking seriously the interaction between their material and metaphorical politics is one of the primary tasks of this chapter.

While Athenry was left at square one, with no incoming development to speak of and Apple still in financial control of the approved Derrydonnell Woods site, discussion across the country shifted focus to the present and future environmental effects of the "data center boom." The environmental externalities of media technologies have been the focus of considerable debate globally in recent years, especially in terms of the export of extractive activities—and their risk—to peripheral zones on national and world scales (see Arboleda 2020; Brodie 2023; Childs 2022). Data centers are merely one part of a growing technological assemblage, global in scale, supporting atmospheric and ubiquitous computing (Gabrys 2016) and "smart" technologies (Sadowski and Pasquale 2015), or "the Fourth

Industrial Revolution," as digitalization has been influentially branded by the World Economic Forum (Schwab 2015). The metaphors and cultural mechanisms by which these occlusive geographies play out have been the subject of a growing number of studies on internet and media infrastructures (Amoore 2018; Burrell 2019; Hogan 2015a, 2015b; Holt and Vonderau 2015; Johnson 2019; Velkova 2020; Vonderau 2019), to which this book contributes. But the "immaterial" frontiers of consumer and business software and "smart" urbanization, where users are sold a wispy, atmospheric "cloud" and corporate planners imagine terra nullius outside of urban spaces to power urban futures, cross through existing and complex rural economies and ecologies. Just like the mines, as sociologist Timothy Mitchell argues, which fired the carbon-fueled urbanism of colonial modernity (2011), the infrastructures powering the digital economy are found outside of cities, in peripheral and colonized zones of extraction, and are navigating through environmental, cultural, and labor relations in their planning, construction, and continuous operation.

In Ireland, these externalities have been debated in terms of how data centers are and will continue to affect energy consumption (see Bresnihan and Brodie 2021; Carroll 2020), and by extension Ireland's climate strategies and targets. Early reports on the data center boom, which occurred in the immediate aftermath of the financial crisis (beginning, roughly, in the early 2010s), were focused on the attraction of Ireland's "cool climate." Cooling is crucial for these infrastructures, which, as halls full of computers constantly humming with the data of our collective business, consumer, and state internet traffic, generate tremendous amounts of heat. Most data centers in Ireland rely on cooling systems that need both cool outside air and water circulation. The cool, wet, windy climate saves (still carbon-intensive) energy and increases efficiency by taking advantage of these conditions. These controlled microclimates thus use local climate conditions while affecting them in collateral ways, particularly via waste heat, increased energy use, and resultant carbon emissions (McLaughlin 2015).

But as the number of data centers grew, so did energy usage in the country, rising steadily in the late 2010s (SONI and EirGrid 2019, 24). While energy demand in most sectors, especially the average household, has remained constant, demand from "data centers and other large energy users" has soared since 2015, rising 144 percent between then and 2020 (Central Statistics Office 2022). In the 2010s, many would reference Greenpeace's early statistic that "if the cloud were a country it would be the fifth largest energy consumer in the world" (Maxwell 2014). In Ireland alone, 2017 reports suggested that by 2026, 15 percent of the country's energy use will go toward data centers (Bodkin 2017), a number that ballooned to 20 percent on world scale projections (Climate Home News 2017). By 2020, estimates doubled: EirGrid projected that by 2028, data centers would make up 29 percent of Ireland's energy use (Carroll 2020).[6] An Amazon Web Services (AWS) data center campus in Mulhuddart, Dublin, uses the same power as a small Irish city, up to 4.4 percent of the country's energy capacity in a 2026 projection (O'Donoghue 2018). The Athenry data center campus in total, according to Supreme Court proceedings in Galway, could have strained the national grid for 5–8 percent of its capacity *daily*.

Whatever the exact numbers, these so-called externalities of digital media business, beyond being unsustainable at the level of supply, will continue to have adverse climate impacts if the national grid remains powered primarily by fossil fuels (53.9 percent of total input natural gas in 2021). While tech companies claimed that they would "provide" and use 100 percent renewable energy within the next several years to accommodate state and supranational energy agreements, an assertion that activists and courts alike often find disputable, the method by which they do so remains to buy renewable capacity in exclusive contracts with the providers (see Bresnihan and Brodie 2021). While from a supply and demand perspective, this adds supply to the grid, it is earmarked for the immediate and privileged supply of data centers.[7] Finally, wherever the energy comes from on a normal day, every data center also needs on-site backup energy supplies in case of grid failure. Most data centers have a supply of continually charged batteries that can power the facility for thirty minutes or so until a new power source is operational, usually on-site gas-powered generators. If the lights go out, the data center still runs, on carbon-heavy power.

While the specificities of these energy politics are urgent, they are not the main focus of this chapter (see Bresnihan and Brodie 2021, 2023; Brodie 2024b). Rather, what this chapter will articulate are the ways in which these climate impacts were being imaginatively offset during the "recovery" by the generation of a different kind of climate: a "business climate," entangled with the natural climate, through which value can be extracted through infrastructural assemblages of data, finance, and energy. Ireland's low tax rate (especially for research and development and intellectual property), corporate-friendly state policy, English-speaking workforce, and history of facilitating FDI, along with the "cool climate," were all mobilized within corporate and state planning to attract or justify the existence of data centers. This chapter shows how the climatic politics of data centers arose specifically as a response to financial crisis, as the austerity environment, desperate for industrial-scale "jobs," was capitalized upon by the tech industry as a place to extract value.

In research on the data center industry in Ireland, I attended corporate conferences, networking events, and Supreme Court proceedings. I encountered people who owned or wanted to see data centers in their hometowns; aspiring and failed data center magnates; industry experts and advocates; data realtors; divided and bitter local communities and individuals; besieged environmental activists; pro–data center state and semi-state officials; curious and intrepid researchers; apprehensive and confused security guards; and even some of the relatively few workers within these black-boxed infrastructures. Access to data center sites was difficult, in terms of both physical location—usually situated on the suburban peripheries of Dublin along the T50 fiber-optic cable route encircling the city and only accessible by car or, in my case, public transit and long walks—and bordering fences with high levels of security. While communities, housing, and ecologies were often visibly left stranded within development zones or pushed out entirely (figure 4.2), many data centers coexisted in uncanny contrast with surrounding environments. The drive for greater connectivity in fact disconnected the spaces of investment from the communities and ecosystems that once existed there, whether leaving them behind or exposing stark inequalities of global capitalism's structural turbulence. Visiting these sites across time provided perspective as to the scale of spatial upheaval required for these and other sites of FDI-driven industry, fueled by the world economy's uneven flows. But the scale and seductiveness of their promise is undeniable. This chapter uses accumulated knowledge and experiences gathered across this fieldwork and planning research to theorize the territorial role of data centers within the Irish state's cooperation with multinational corporations in and through the country. These policies, which enact the FDI-led economic common sense of Irish industrial development, served to naturalize the role of big tech within the country's social, political, and infrastructural futures.

Data Center Country

To map the geographies of data centers in Ireland, we should return to where we left off in chapter 2—Dublin. The epicenter of the data center "boom" is Dublin's sprawling suburbs, a blast radius from the country's

4.2 A boarded-up house in sight of Google's data center in Grange Castle Business Park South, winter 2019. Photo by Patrick Brodie.

FDI-led economy encircling the city along the T50 fiber-optic cable route and the transport, energy, and industrial infrastructure with which it is bundled (notably along the M50 ring road). Private data infrastructure clusters around the projects and infrastructural resources provided by the state or other corporate presences, and this is especially true around Dublin's "databelt" where most of these data centers have concentrated at the edges of the city to take advantage of the available land and infrastructure (McLaughlin 2015, 198). This formation emerged since the first hyperscale data centers arrived in the late 2000s, at the height of the financial crisis.

As geographer Martin Danyluk argues through what he calls the "logistical fix," capitalism's inherent and contradictory turbulence is reflected in the transformative global distribution of logistical infrastructures like data centers, managerial technologies that shape the spatial distribution of industry across the world's territories and supply chains (2018). Irish architecture scholar John McLaughlin demonstrates that this formation reflects the attractiveness of the country's tax climate but also the clustering of data centers around other critical and circulatory infrastructures (2015). The infrastructural "bundling" is key to identifying the strategic routes through which these forms of public/private partnership operate. Just as logistics companies require public roads and railways as much as natural and man-made waterways to operate, private high-tech companies require and instrumentalize the energy and fiber-optic resources of territories as much as its labor resources and, in this case, climate conditions (figure 4.3).

Infrastructure, as media scholar Shannon Mattern tells us, is "path dependent" (2017, xxviii), meaning that it follows routes laid down by existing physical networks, reflecting their histories and geographies of power and influence. As the CEO of Cork Internet Xchange (CIX) told me, the site of his relatively small colocation data center in Hollyhill Industrial Estate was in part determined by its proximity to Apple's Cork headquarters, as the concentration of fiber-optic cables and other internet infrastructure was already piped up the hill on which CIX operated. Media scholar Mél Hogan notes that this occurs because "Large corporations also pay for various infrastructures that are intended to help other server-based companies 'set up shop' nearby. So while communications technologies have long been privatized, these internet infrastructures are increasingly entangled in market logics that are making internet flows a utility to manage, like electricity and water" (2015b, 4). The internet access of the nearby housing estates, which stretch for miles surrounding CIX, would be pay for subscription like everywhere else, while CIX's mechanical sheds extract value from internet

4.3 Bundling and path-dependence. At Citywest Business Park in spring 2019, home of several data centers, bundling of electrical and transport routes at the Luas station is in evidence. Photo by Patrick Brodie.

traffic all over the country and beyond. While a "utility to manage," access to it is differential and profit-oriented. The differential access across private companies, facilitated by the state, has a compounding effect on the industry. Nonetheless, internet infrastructure, as supposedly ambient and public architecture for the circulation of information (see Larkin 2013), is treated as a naturally occurring part of the Irish environment from which other companies can extract value. Companies simply need to get planning permission and a connection license, and plug in.

As Holt and Vonderau point out, the design and location of a data center mimic the design of something like a warehouse (2015, 72), or "sheds," as architects Gary A. Boyd and John McLaughlin describe them (2015), reflecting the modular logistical routes that make up the actually existing contemporary geography of industry, layered on top of existing infrastructures, flexible and adaptive to changing conditions. Data sheds are rapidly built, faceless architectures, futuristic in the very mundaneness of their scale (figure 4.4), appearing purpose-built for some other future purpose, often constructed on spaces of former industry. Technological obsolescence and ruin are hard-coded into their territorial DNA and the corporate visions they represent.

However, too-intently mapping infrastructures' path-dependence, logistical organization, and "smoothness" simplifies the complex territorial politics that determine these formations in contemporary capitalism. These flexible, elastic rhythms of the global economy are operational to what Tsing has called "supply chain capitalism" (2009), the contemporary transnational arrangement of trade and power by which diverse life and labor enrich the supply chain while ensuring structural subjugation to extractive capital. But central to the thesis of supply-chain capitalism, and studies of logistical infrastructures that have mapped and analyzed these global circulations (Cowen 2014; Mezzadra and Neilson 2013b, 2019; Ong 2006; Tsing 2005), are the territorial and environmental formations "stitched together" (Rossiter 2016; Starosielski 2019) by the governing and management of logistical technologies. Whether speaking of national borders or the separated and strategically zoned infrastructures and office parks encircling today's cities, zoning and bordering are used to separate workforces and administer territory for the greatest exploitation by capital (Mezzadra and Neilson 2013a; Ong 2006). As borders have multiplied across geopolitical contexts as well as across everyday life in the age of finance capital, tech companies plan infrastructures not necessarily through the smoothest path. Rather, they navigate operations across social, cultural,

4.4 Faceless AWS data center near Tallaght in spring 2019. Photo by Patrick Brodie.

political, and environmental avenues of friction and resistance. If successful in doing so, their functional operation and power of extraction are strengthened and expanded.

The data economy is continuously navigating across different spaces via networks, and these spatial differentiations are part of how the supply chain continues to function. As small and modular edge data centers are built throughout cities and their "smart" environments to accommodate 5G (Enterprise Ireland 2017), larger hyperscale and colocation centers continue to grow in the outskirts of these same cities. On multiple occasions, I took the 151 bus from the Coombe in Dublin 8, where I lived during my fieldwork, out to Grange Castle Business Park, established in the early 2000s as a partnership between the Industrial Development Authority (IDA) and the South Dublin County Council. The sprawling industrial region within which it sits, just west of Clondalkin and situated along the historic Grand Canal and among the ruins of medieval Grange Castle,[8] hosts dozens of multinational companies—and especially hyperscale and large colocation data centers from some of the biggest-name cloud providers (including AWS, Digital Realty, Equinix, Interxion, Google, and Microsoft). Roughly across the way is Google's flagship Dublin data center in Grange Castle Business Park South, and a neighboring Digital Realty data center in Profile Park, which infrastructurally positioned itself as a designated hub for data centers—like other IDA parks (see Brodie 2021; B. O'Halloran 2017).

During one of my visits to the area, my destination was Interxion. Upon arrival, I was greeted with typical security: After procuring an appointment weeks earlier, I had to announce myself at a separated gate, be buzzed into the parking lot, then buzzed in again at the door, where I had to provide government ID, sign in, and wait for a representative. When this employee finally arrived in the glossy, empty coworking waiting room, he took me on a brief tour of the computing facility, where I once again had to be let in, this time by biometrics. This was a familiar experience for me by this point in my research. Like other such spaces I had visited, after security, the inside was largely unremarkable for someone without engineering expertise: Few personnel were seen, hallways and workstations were squeaky clean, rows of humming server stacks made up most of the rooms we entered, backup energy generation and battery power were essential spines of the facility, an electrical substation sat just offsite, and as usual, they were expanding their data halls into a new building next door.

Upon concluding our technical tour, we sat down in one of the conference rooms, with a view of the fountain pond near the entrance. Speaking with him about not only Interxion, which hosted several other facilities across the Dublin region, but about the broader data center economy of Ireland, the representative (who was born and raised in Ireland) told me that the country was uniquely suited for data center development. This went beyond typical justifications, however, like tax incentives and climate factors: He argued that there was something about Irish *culture* that made it a good place to construct and operate a data center. As we spoke, and then later as I wandered these inhuman industrial landscapes on my way back to the bus stop, I was reminded of how he greeted me upon my arrival: "Welcome to data center country" (see Brodie 2021). While he spoke of this data center–laden region encircling Dublin, this jokey statement echoes the state's vision of the country as a home for these energy-hungry infrastructures.

Ireland's influx of data center development by some of the largest tech companies and data colocation providers in the world has been a subject of considerable interest and debate, both in Ireland and in the data center industry (Carroll 2020; Jones 2014; McDonald 2012; Silicon Republic 2016). Ireland in the 2010s was the most concentrated data center hub in Europe (Savvas 2018), and most of this development was centered on the Dublin region, ramping up since the 2007–8 financial crisis (see IDA Ireland 2018, 29). In an environment of extensive job loss and austerity, like the wider "creative economy" in chapter 2, data centers' massive investments were positioned in public policy and discourses as tools for "recovery," central to the imaginary of the "Celtic Phoenix." To provide a brief recap of the timeline in chapter 1, Microsoft and AWS planned or built data centers as early as 2007; Facebook built its first data center in 2016; Google expanded into its Grange Castle facility in 2016; and, finally, Apple tried but failed to build a data center in Athenry, County Galway, between 2015 and 2019. Microsoft, Google, Facebook, and especially AWS, the so-called hyperscalers, have extended facilities, the latter with over a dozen (by my count) data center sheds across multiple areas of Dublin as of 2019, which makes up its EU-West-1 Availability Zone.[9]

These companies coexist among dozens of other multinational colocation data center providers (like Interxion, Equinix, Keppel, and Digital Realty) hosting expansive operations, established roughly across the same timeline. Still more companies experimented with edge, modular, and other schemes like underwater cooling facilities to deal with availability, space, and heating/cooling concerns. Smaller operations like CIX (Cork) and data management and services companies like Sleepless (Galway) filled in gaps and service this multinational supply chain. Analyzing public materials, websites, industry documents, planning permissions, and court cases from this period, one finds repeated and all-too-familiar claims as to the benefits for Ireland or hosting data: its open business environment and friendly tax climate; a highly educated and English-speaking workforce; multinational companies for industrial partnerships; quality digital and electrical infrastructure; cool and damp weather conditions. However, this rosy, top-down view of Irish space, industry, and its planning systems glosses over the messy and conflicting interests that each data center project must inevitably negotiate. Objections became more obvious as the number of data centers grew: Construction tended to be short-term, jobs in the completed data sheds were limited and often, in the case of colocation data centers, employees were contracted or shipped in from abroad by companies renting server space. Energy use continued to balloon. Water supply concerns, not associated with wet, rainy Ireland, were reported in Dublin, drawing attention to data centers' resource use beyond only energy (Woods 2020; also Hogan 2015a). Reports of "saturation" (Hogan 2021) and potential energy shortages in Dublin (see IWEA 2020, 2) led to early calls for a moratorium on their development[10]—and suggestions that data center providers should start looking outside of Dublin for expansion.

Companies had already been planning for this by 2019 when I finished my fieldwork—Apple (Athenry), AWS (Drogheda), T5 (Cork), Echelon (Arklow), and Atlantic Hub (Letterkenny/Derry), among others, were already looking outside of the immediate Dublin fold to gain an Irish foothold in places with less grid strain, closer to the renewable energy generation that they will need to remain compliant with Irish and EU policies and targets. Proposed data center projects in Killala (County Mayo), Ennis (County Clare), and the Bord na Móna Energy Park (across Counties Offaly, Westmeath, and Meath) have claimed similar reasons to attract investment, promoting infrastructural availability and on-site energy (Bresnihan and Brodie 2021, 2023). Like other forms of speculative property development, aspirational data center proposals were put forward regularly across the country. For years, there were few caveats as to energy costs. Places beyond Dublin wanted to be a new "Silicon Valley," in the case of Athenry (Siggins 2017), or a new "goldmine," as a planned €450 million data center in Ennis heralded (McMahon 2019). The "gold rush," as the metaphorical resource politics engaged in by data "mining" proclaims new automated geographies of extraction, was on. Sometimes, the state succeeded in pushing a project through with little resistance; other times, projects hit unexpected blockages and disputes, or foundered in the courts, usually around environmental objections. For example, Nautilus Data Technologies, a start-up, for several years attempted to secure permission to build an experimental underwater data center in the Limerick port. Residents and business owners objected, arguing that heated wastewater would damage the marine ecosystem. After a few years, state and capital's support prevailed, and the project went ahead (Judge 2019). Such is the nature of extractive capital and its modes of territorial expansion—while capital is efficient, and the state paves the way, politics and environments remain unruly.

Despite largely supportive policy frameworks, there were frictions even within private, semi-state, and state organizations in charge of managing data-center planning and construction. From 2017 to 2019, many began to alternately sound the alarm about data centers' energy usage while also arguing that they are an essential part of the state's growth strategy (Enterprise Ireland 2017; Government of Ireland 2018a; IDA Ireland 2018; IWEA 2020; SEAI 2017). These interests appeared to be in conflict, but reading through these documents, they were somehow posed as part and parcel: Data centers did and would continue to use *astronomical* amounts of energy in Ireland (see Bresnihan and Brodie 2021). However, the technical challenges of these energy demands would be met with methods of "green" growth and an expansion of renewable energy capacity. For example, while the Sustainable Energy Authority of Ireland (SEAI) (in collaboration with civil society booster Host in Ireland and Bitpower Energy Solutions) said that energy supply would become more difficult, it also bought into the wider need for data centers in the Irish economy in relation to global trends and data-driven enterprise: "Investment in data centres is significant, and the data industry supports many jobs across the economy. There are also opportunities in the content of the data. We are only at the beginning of the digital age, and Ireland needs to be ready to leverage future trends in data. . . . Collaboration between the data centre operators, the state utilities, renewable developers, researchers, state agencies, and local authorities will be key to unlocking future opportunities" (SEAI 2017, 29). Such reports speak to the concerted and integrated effort across public and private sectors to secure growth via strategic industries like data centers. The investment offered by these infrastructures was too significant to merely scoff at, and the imaginaries of tech-driven economic prosperity—even "ripple effects"—were seductive for many people across Ireland, especially in the post-crisis environment. However, industrial change and the basic realities of low job numbers at these sites risked gambling rural and wider developmental futures on the speculative and profit-driven business prospects of tech companies searching for the most favorable conditions. If Ireland ceased to provide those conditions, they could leave.

Legislative change to facilitate data centers has been widespread. As the Fine Gael Government's strategy documents indicated, there are several pillars to their incentivization, including changes to planning and energy policy and data centers' inclusion in the Project Ireland 2040 spatial planning framework (Government of Ireland 2018a, 3). Taoiseach Leo Varadkar announced in 2017 a proposed amendment of the Planning and Development Act of 2000, an extension of the 2006 amendment on Strategic Infrastructure, which paved the way for strategic development zone (SDZ) designations, moving to "allow data centres and other key IT infrastructure to be included in the criteria for access to the Strategic Development planning procedure" (Government of Ireland 2017). This classification as critical infrastructure would be leveraged to fast-track development and "enable the planning process to work more smoothly" (quoted in Finn 2017). The amendment, according to Varadkar, would "allow them to skip a whole state [sic] of the planning step into the future and in addition to that we're going to develop a very clear policy around data centres, where they should be located and how energy should be provided for them" (quoted in Lynch 2017). While companies would still have to receive planning permission at the state level, they would no longer be required to follow local planning procedures, lodging their plans immediately with the statewide, quasi-judicial An Bord Pleanála. As outlined in chapter 2, like designations of "critical infrastructure," these "strategic development" procedures remove local considerations from these large-scale projects (see Byrne 2016), effectively attempting to bulldoze local political mobilization reflected in the Athenry saga. The messy, conflicting interests in the west of Ireland constituted a barrier to be overcome by deregulation.

To Seán, the planned strategic infrastructure amendment (which never went ahead in that form) was an empty gesture even in itself: Objectors, he believed, would still have slowed the process even if the first round of appeals had been circumvented by the streamlined law. These aggressive moves to pave the way for FDI thus demonstrated not only the state's aversion to objection, but their strategic reliance on FDI despite the technical and political challenges presented by digital growth. Varadkar attempted to steady the ship and ward off corporate concern after Apple's withdrawal, saying that such objection is "not the norm. There are lots of data centres all over Ireland and they get through the planning process with relative ease so I don't think this delay in Athenry because of the courts and because of the planning process is typical" (quoted in Lynch 2017). The courts seemed just as eager to alleviate concern. In addition to the strategic infrastructure amendment proposal, the Supreme Court's 2019 ruling decided in favor of Apple's right to not provide a full environmental impact assessment (EIA) for their plan, where they had only assessed the impact of one data center shed as opposed to the proposed eight. This ruling came in spite of Apple's withdrawal, and set a precedent meaning that other companies can follow Apple's route of underplaying the environmental and energy costs of their future plans.

Through these political maneuvers, the state enabled a system of infrastructural facilitation wherein data centers became inevitable factors in Ireland's economic future. It also, more alarmingly, facilitated a coalescing unity of logistical governance, through the territorial technology of the data center: By removing or attempting to remove barriers to development by preemptively subduing and reducing avenues for dissent and alternative political mobilization, the state has tried to outflank public objection before it has a chance to form. As Ned Rossiter articulates in terms of what he calls "the logistical state" of contemporary capitalist territorial governance, "If logistical operations are central to the emergence of new forms of sovereign power, then a media theory of logistics lends analytical traction to the collective work required to diagnose and critique contemporary regimes of rule" (2016, 4). The encircling barriers to democratic participation in these planning processes came about as a direct result of a state strategy entirely devoted to FDI as a model of economic development, intensified by the desperate logics of austerity. Central to these emergent forms of politics and rule in contemporary logistical capitalism, administered by tech corporations, their infrastructures, and their state partners, are forms of security and strategic visibility, which interlock with the politics of visuality put forth by the Irish state and its economic strategies.

Visibility and Infrastructural Climatics

As chapter 1 argues in detail, the privatization of infrastructure is constantly in action, and often the result of supranational debt mechanisms and dependence. FDI is a strategy to draw capital into strategic geographical areas and sectors of the economy, including culture, public services, and infrastructure. The blurred boundaries of public and private provision within supposedly "public" infrastructure means that, as elsewhere in the neoliberal West, state subjects are increasingly treated as consumers, paying customers rather than citizens deserving care, with the politics of "access" often replacing welfare and care (see Ranchordas 2018). This wider infrastructural formation acts in support of the dominant flows of data and capital in the tech world. In particular, many multinational tech corporations in the 2010s were concentrated on shoring up the streaming space and content production. Amazon Prime, Apple TV, Disney+, and Netflix (the latter two contract their server space) require expansive data center facilities to accommodate and distribute the streaming traffic of their millions of users worldwide.[11] But in the same data sheds, there exist consumer data, business records, government files, and personal cloud storage. These convergences of technology and content, production and distribution, governance and consumption, while part of a massive infrastructural apparatus spanning territories, cables, wires, nodes, and a variety of important sites and ecologies, are tangled within the territorial technology of the data center. Within this environment, data centers are given exceptional purchase and promoted as public projects bringing green and data-driven prosperity, as Athenry and community support for other projects demonstrate. They are seen as essential and integral infrastructures to the state's economic future, despite their largely extractive presence and their unmanned operations.[12]

Contrary to the public services that they are promoted to offer and the strategic role they play in state strategy, data centers are notoriously guarded or strategic about when and to whom they make their operations visible. Transparency around these infrastructures remains a major point of contention, as the idea of the cloud fundamentally occludes where our personal data is stored and what power governs it (Holt and Vonderau 2015, 75). For example, AWS is deeply secretive about their operations, to the extent that a leak of their "infrastructure regions" by Wikileaks in 2018 was enough to provoke an international scandal. Scholar and artist Paul O'Neill made it part of his artistic practice to bring "tour groups" out to AWS data centers surrounding Dublin, allowing visitors (mostly artists, activists, and academics) to witness their continued operations and expansion (see Humphries and Kapila 2019; O'Neill 2018, 2019). These visits involved walking through these inhuman landscapes, in often pedestrian-averse areas clustered within suburban enterprise zones and surrounded by other logistics centers, manufacturing operations, and warehouse spaces. Referencing and in conversation with infrastructure scholar Ingrid Burrington's longtime practice of bringing groups on infrastructural tours of New York (2016) and Mattern's concept of "infrastructural tourism" (2013), O'Neill built an experience out of the facilities' amplified security, even provoking the security guards—from the safety of public property, outside of the data center fence—to engage, usually through loudspeakers installed on the outside of the data centers. Security guards, when asked, would not tell you who owned the buildings, despite the publicly available planning documentation (although even this was sometimes occlusive, using misleading subsidiary acronyms such as ADSIL, Amazon Data Services Ireland Ltd.).

In my own experiences visiting AWS (with O'Neill and others) and other data center campuses unannounced, encounters with security were inevitable, especially when stepping foot on any manner of private property. Once in summer 2017, after taking a bus out to the then far less developed Grange Castle Business Park, walking along a narrow road with no sidewalk from the bus stop to reach an official entrance, I traversed the poorly available public walking infrastructure of the area and finally approached the gate of the new Google data center. Taking photos the whole way, I stepped inside the gate and was immediately apprehended. Private Google security guards took my passport, asked for my information and affiliation, and called their supervisor. In all, I was held there for roughly twenty minutes, with a few tense moments before their tone changed to one of more curiosity than distrust. I had a similar, if less hostile, experience with a colleague in 2019. We entered the front door of an unmarked Equinix data center in Citywest Business Park, to confirm who owned it. The security guard, in shock that we had just walked in, asked us to leave immediately. These highly secure infrastructural (and labor) accounts, however, are not universal. Next door to Equinix in Citywest, my colleague and I approached an unmarked data center owned by Keppel, a Singaporean conglomerate, and the security guard—a contractor, based on his uniform—greeted us cheerfully: "Here for Keppel?"

Other companies offer public-facing visibility to control the narrative. Google presents itself as a public space, complete with a mural painted on the side of its main shed, but visible only from a safe distance (figure 4.5). Apple, similarly, performed a good deal of community visibility work in Athenry, donating iPads to a nearby school and promising community walking trails, before the project was canceled. Facebook gave a publicized tour of their facility in Clonee to state broadcaster RTÉ (Goodbody 2018). Like their operations in Prineville, Oregon (see Burrell 2020) and Luleå,

4.5 Google's Grange Castle mural , all clouds and wind, in spring 2019. Photo by Patrick Brodie.

Sweden (see Vonderau 2019), Facebook tried to build consent with *strategic* friendly visibility. However, as anthropologist Asta Vonderau notes, the data center in Luleå remained mostly locked away, its visibility controlled by a PR firm in Stockholm, despite ongoing community support (707).

As chapter 2 demonstrates, the visual culture of development represents these financial and logistical modes of seeing and governing the future by erasing existing communities, labor, and political dissent. Parallel to how these are rendered occluded practices and geographies, there is a politics of (in)visibility to securitizing of resources and their circulation. As media scholar Nicole Starosielski argues in terms of undersea telecommunications cables, such infrastructure, always fuzzily public and private and existing across jurisdictions, has tended to be invisible to the public eye (2015). By surfacing localized histories and social relations around such infrastructure, Starosielski activates different ways of seeing them and their geographies. However, that is not to say that corporate and state strategies do not tend to manage these visibilities, making any activity of envisioning and mapping alternative politics a difficult endeavor. Only by controlling and managing where and how the public (or investors) become aware of data centers can the cloud truly recede into the naturalized ambience of global life.

In relation to these politics of transparency and visuality, Holt and Vonderau propose seeing popular visualizations and corporate propaganda around data centers, along with the physical spaces, policies, and agreements, as part of their infrastructural operation and necessary to understanding the public/private imbrications of the cloud's built environments (2015, 74). Resonant with discussions of corporate visualizations of future urban space in chapter 2, data-center plans project an imagined, digitized future. In these images, problems and turbulence are nowhere, despite the values assigned to manage the messy contingencies of infrastructures and environments through these logistical technologies. Speaking of visualizations of data centers, Holt and Vonderau argue that "such images tell us about affordances and constraints turned into pipes and cables, about in-built political values and the ways the engineering of artifacts come close to engineering via law, rhetoric, and commerce. And the images also testify to the constant struggles over standards and policies intrinsic to the network economy" (2015, 74). As suggested early in this book, the visual culture of such images projects ideal infrastructures through which frictionless imaginaries can become operational, despite often fraught planning processes, political dissent, and ecological friction that characterize their development. These physical infrastructures are not simply those that are built, but also the surrounding visual culture, which circulates and contributes to the collective imaginary of a given project.

Such images also promote the "green" vision of technology, a clear-skied digital future driven by the clean, human-free infrastructures of cloud computing. Visualizations of data centers, through corporate promotional material and press releases shown to the general public on industry and journalistic websites, often focus on the "expanse of sky and land surrounding the buildings. In effect, the data centers are visible but rendered practically inconsequential by the surrounding spectacle of natural vistas and wide-open spaces" (Holt and Vonderau 2015, 76; see also Carruth 2014). Data centers are energy vacuums, and yet these images do not manifest the surrounding infrastructural or environmental conditions, unless supporting a green image (for example, on-site wind generation). Companies vie for position in a market increasingly geared toward "green" alternatives, which governments (state and supranational) incentivize to mitigate damage while encouraging growth. Industry and company websites inveterately preach corporate dogmas of energy efficiency and green energy that motivate the rhetorics of sustainability and climate responsibility. Environmental scholar Allison Carruth describes the ways in which industry propaganda around "the green cloud image often serves to greenwash both network infrastructure and corporate America," because, "whether lauded as a silver bullet for corporate sustainability or exposed as much dirtier than we think, the medium works . . . to simplify the 'large, complicated, [and] inaccessible' infrastructure that moves data around the world" (2014, 349). These facilitators of secure circulation require intersecting promotional and infrastructural tactics, as they need not only to be perceived as sustainable but to be embedded in the transformation to lower-carbon economies (Brodie 2024b). Rendering the cloud "green" through these strategic visibility tactics paints corporations as "sustainable" elements of unstoppable forward movement, naturalizing their relationship of care toward future planetary ecologies (see Hogan 2018a).

Crucially, and betraying the few jobs actually created by data centers, the buildings within these "green" visualizations are part of larger, usually pristine, natural landscapes (figure 4.6), devoid of people. Returning to the politics of urban and rural development in Ireland discussed in chapter 1, these images conjure the "greenfields" imagined by state propaganda to bring capital into rural spaces, as literary scholar Luke Gibbons has identified in early economic development materials promoted by the IDA

(1996). Like the aspirational promotional spaces of the city expressed in chapter 2, visions of development—or, perhaps, of technological urbanization—promoted by state and corporate planning are unevenly transferred across the country and demonstrate the strategic (em)place-ment of data infrastructures and their demographic imaginations through regional variations. Similar to the architectural renderings of planning proj-ects in chapter 2, the "place atmospheres" (Melhuish, Degen, and Rose 2016) generated within images of data centers, for example, assume a par-ticular view of territory: terra nullius, formed against green landscapes, clean, devoid of people. Such atmospheres, what infrastructure scholar A. R. E. Taylor refers to as the "technological wilderness" of data center imaginaries (2019; also Lally 2021), enact logistical violence by imagining an empty landscape for capital to colonize. However, rather than necessar-ily external imaginations, their visions in and of Ireland are endorsed by the state, profoundly seductive in their visuality, imagining these "empty" spaces as potentially green, technologized, future spaces for sustainable and automated prosperity in places often outside of the fold of these corporate visions of urbanization.

However, to push against these visions, we must pay attention to their on-the-ground politics, and recenter these peripheralized rural and semi-urban areas within our analysis of such systems. By paying close attention to these politics, we can short-circuit and complicate top-down visions of corporate urbanization. Focusing intently on rural "frontiers"—the "sys-temic edge" of technological urbanization, as Mezzadra and Neilson refer to such sites of development (2019, 138)—can decenter centrifugal analy-ses in order to locate much more complex and multisited entanglements, the many swirling centripetal intensities produced across global systems. Google's data center in Grange Castle Business Park South, for example, as recently as 2017, used to be surrounded by fields and small estates, just on the edge of Dublin's suburban sprawl. In a very visible way, this sprawl caught up: Roads have been developed, infrastructure expanded, and in the process, what was once farmland has been claimed for development. Compulsory purchase orders from the state, issued long before I had even visited, and the continuing expansion of industrial and data center oper-ations in the area, had left houses and properties stranded by the time of my last visit in 2019 (see figures 4.7 and 4.8).

These processes are often far more intensive to these places—and with active, albeit frequently dismissed and disempowered participation by local people—than such extractive frontier imaginaries leave space for. This is

4.6 Digital rendering of Apple's proposed data center in Athenry (Wuerthele 2018).

4.7 Microsoft's data center campus in Grange Castle Business Park, viewed from a new local road and bike path, spring 2019. Photo by Patrick Brodie.

4.8 In-progress suburbanization, with an abandoned farmhouse and the roof of Google's data center in Grange Castle Business Park South barely visible at center right, also spring 2019. Photo by Patrick Brodie.

especially true of Athenry, where familiar infrastructural and planning arguments as a strategic frontier for development of crowded Galway city were fed by profound local support and political mobilizing by the community, attempting to facilitate large-scale FDI by an enormously wealthy transnational corporation.

Extracting from Climate

In 2015, Apple first announced its plans to build a data center at Derrydonnell Woods. Referred to as a "greenfield" site on the outskirts of Athenry, "neither Coillte nor the IDA would divulge details of the land transfer to Apple" (Irish Times 2016).[13] Twenty-five sites in Galway were inspected with the IDA, but some "were immediately dismissed because of poor roads and broadband, or else the sites were too small to cope with a 15-year plan to build eight data centres in one location" (Newenham 2016). Apple officials stated that Derrydonnell was "uniquely attractive" (Newenham 2016) and, as Seán told me, the natural cooling offered by surrounding woodlands would apparently supplement Ireland's already cool climate. Athenry is also located near the junction of M6 and M17/M18 motorways, a fiber-optic cable route, an IDA site, and a commuter railway connection, which would make the forest cooling appear secondary (like the cooling arguments in Ireland more broadly). On the surface, Athenry seemed to merely be a strategic "dot on the map," among a bundle of developments and infrastructure (figure 4.9).

However, from the beginning, local politics were complicated. Apple consulted with the community, who formed the Athenry for Apple group in support; state officials representing the region advocated strongly for the project; and, among the overwhelming excitement, several environmental objections were lodged at the initial planning stages. These objections held up the process until October 2017, when, after a series of juridical delays— including the arrival of climate change–influenced Hurricane Ophelia on Ireland's shores (Carolan 2017b)—a Court of Appeals dismissed the objections. However, the court agreed that Apple "had noted no direct renewable energy connections or renewable energy projects" in the plans for the data center, nor had the company "clearly shown power to the data centre would be from 100 per cent renewable resources" (Carolan 2017a). Despite this, it was determined by the courts that "the potential employment and regional development benefits of the centre outweighed potential adverse climate impacts" (Carolan 2017a). This admission that the climate impacts

4.9 Derrydonnell Woods outside of Athenry, a strategic "dot on the map," summer 2018. Photo by Patrick Brodie.

were secondary to the whims of capital investment, as well as Varadkar and Apple CEO Tim Cook's close relationship in shirking collection on Apple's €13.1 billion in back taxes owed in the country, put Ireland and Apple on a legislative crash course with the EU. In November 2017, the Taoiseach's diplomatic meeting at Apple's headquarters in California—a geopolitics of state/corporate partnership—cast doubt on the project's future despite its court approval. Varadkar announced the new uncertainty, regretting that the appeals process was outside of Apple's control.

In December 2017, the objectors lodged a final appeal with the Supreme Court. The Athenry for Apple group expressed regret for continued dissent, at a local and national level (RTÉ News 2017). On May 10, 2018, after three years of negotiations and delays, Apple announced that it would no longer be proceeding with the project. This came two weeks after the Supreme Court allowed the principal objectors to continue their environmental appeals, which may have landed Apple on another time-intensive crash course with the EU (Wuerthele 2018). Despite this and ongoing controversies about unpaid taxes, Apple and the Irish state maintain a good relationship, and even after the "Apple hangover," members of the Athenry for Apple group pressed on, continuing to share stories and thoughts in their dedicated Facebook group until at least 2020. While the state adapted to ensure data centers continued to locate in the country, the people of Athenry and East Galway were still at square one, with no incoming capital to speak of. From the community to state levels, there was embarrassment and anxiety about scaring off future investors. Countrywide, cases of planning objection referred to Athenry as a cautionary tale. Thus, in the town especially, unease around this loss of potential investment crystallized into hardened resentment toward the state and objectors. Aspiration turned to bitterness.

Social histories and climates of dissent and protest, as much as support, are encapsulated within what I mean by "climate" in the expanded sense throughout this chapter. In Athenry, and in the west of Ireland more generally, the governmental calculations of state and capital failed to account for what those in the town, and various friends and acquaintances elsewhere, referred to as a "culture of objection," "serial objectors," and "professional agitators"—an atmosphere of dissent commented upon by rural scholars like Hilary Tovey (1993) and Robert Allen (2004) as a productive, vital political condition. In a less positive conversation with East Galway Teachta Dála (TD) Seán Canney, he compared the objections in Athenry to those of the Galway ring road. Traffic in Galway is notoriously congested, with ongoing issues of space and expansion for an otherwise small urban center. However, a ring-road proposal to ease congestion was delayed for years in the Irish courts. The project was eventually brought to the EU, which upheld objectors' environmental conservation concerns about the proposed route, and the project was put on hiatus. As a local Fine Gael councillor complained, "The land is just growing briars, it's scrubland and good for nothing. . . . We seem to have this attitude of object to everything and get nothing done" (quoted in Melia 2013). Canney, among others, maintained that in the case of Apple in Athenry the objectors' use of the appeals process was more about the duration than the substance of objections, advocating strongly for the greenfield developmental promise of multinational infrastructure.

But as we can see in objection as much as in support like that of Athenry for Apple, culture, as much as climate and weather, reacts unpredictably to finance and technology. These sociocultural-environmental elements on the ground are pointed to within the headline about Athenry alluded to earlier, that Apple was "stuck in mud in the fields of Athenry" (Paul 2016), referring to the Irish rebel folk song from the 1970s. Sung by a prisoner being sent to a prison colony for rebelling "against the famine and the crown," the reference to "Fields of Athenry" activates this history of rebellion in spatial as well as temporal conversation with Athenry's appeals for the data center. Like such headlines, residents demonstrated willingness to put current development in historical exchange with Athenry's cultural heritage. Seán contrasted his family's longevity and heritage in the area with the objector Fitzpatrick's, who also lived adjacent to the site but was referred to, with bemused but dismissive sympathy, as a mere "NIMBY" by other people with whom I spoke. While her claims were related to environmental conditions and a sense of intrusion, Seán felt as though a data center would be friendlier to the ecosystem than other proposed industrial plans for the site. Resident concerns for local ecological conditions were factored into supporters' advocacy for the Apple plan, contrary to Daly and Fitzpatrick's environmental objections.

Such culturally and ecologically focused support recalls early discussions of Google's investment in data centers in Ireland, proposed by Irish weather writer Charlie Connelly: "Maybe Ireland will now embrace its climate. Some have tried already, most notably the 19th-century writer William Bulfin from County Offaly, who described the Irish rain as 'a kind of damp poem. It is a soft, apologetic, modest kind of rain, as a rule; and even in its wildest moods it gives you the impression that it is treating you as well as it can under the circumstances.' But [Google's investment in data centers] is probably the first recorded case of anyone planning a move to Ireland because of the weather" (quoted in McDonald 2012). This instrumentalization of the environment to attract transnational capital, and the thickness of entanglements at these sites, demonstrates material and ideological relations captured by the state and capital within projects like Athenry. Seán's family history in the area and his concern for the local environment, to him, seemed to strengthen his claim on the space and the desire to have Apple there. Seán's rooted attachment to Athenry, its environment, and the surrounding region actually led to greater feeling about the necessity for transnational investment on the site.

Recognizing that "climates" are co-constituted by structural forces as much as the everyday is crucial to understanding how the existent conditions and those created (and data generated) become sources of value extraction for capital. Through these environmental engagements with media, pasts, presents, and futures of media materiality and climate politics intertwine through the infrastructural form of the data center. These entanglements, while representing perhaps a "poetics of [media] infrastructure," as suggested by Larkin in his analyses of such intertwining spatial and cultural politics (2013), are not only happening within culture and public discourses. They are also *tools* of association that the state and capital use to naturalize the flow of data and the presence of multinational extractive corporations in the country of Ireland.

The "cloud" and its discourses tend toward the metaphorical and make wispy and "atmospheric" data circulations that are resolutely grounded in the cables, sheds, and server stacks of digital networks and the material energetic processes—from peat to wind and across the processes required to produce, distribute, and access this electricity—that sustain them. In addition, associations of data as the new gold or oil demonstrated the industry's desire to supplant association of old-school resource extraction and "dirty" production with the new, dematerialized productive politics of the cloud. The SEAI's data centers report makes similar associations: "As the 'Connected Planet' becomes a reality, the need has arisen to securely store and forward the 'oxygen' on which this growth is based" (2017, 29). Atmospheric gases are continuously affected by ongoing industrial emissions and the release of carbon from landscapes degraded and extracted for fossil fuels, and data and capital colonize the air with their climatic resource politics. The naturalization by state forces of data as a source of value for extraction is then materialized in the territorial form of the data center through a variety of social, cultural, political, and environmental conditions. We need to take these associations seriously to understand how they exert ideological and political weight on the material processes of data infrastructural construction.

Revealingly, the fact that SEAI employs the atmospheric association of "oxygen" with the "cloud" and its material infrastructures demonstrates a wider metaphorical and circulatory politics of data and capital. Newspapers, magazines, policy, and corporate literature all echo: Ireland's climate reduces the immense cost of cooling data centers. In a revealing article from the early days of the data center boom, a *Guardian* article formulated the appeal in popular terms: "'It's not often that Irish weather is a cause for praise, but the temperate climate was very significant in choosing Ireland as a location for this data centre,' says Dan Costello, Google's global data centre operations officer. The group has managed to reduce the amount of energy it uses worldwide to cool down its data systems to just 12% of its energy bill. 'It's not quite as simple as just opening the windows, but it's pretty close'" (McDonald 2012). The tying of climate to culture—in the form of national pride and praise—here repeats the return of certain essentialisms of Irish space (and weather) in the service of transnational capital, a contemporary iteration of what Yuriko Furuhata calls the "climate determinism" of colonial geography and its deployment through particular national spaces (2022). This atmospherics of capital was central to the landscape of culture and technology in post–financial crisis Ireland, but has a much longer history in the way that the imaginary of Ireland as a wild green frontier has facilitated the operations of mostly US FDI for decades.

Thus, the above segment's familiar tool of economic development is freely read through public, corporate, and state discourses: "Ireland has been able to attract these world-famous corporations despite the depth of its financial and economic crisis, due to the lobbying work of the country's Industrial Development Authority; a highly educated, young, English-speaking workforce; and, crucially, the Republic's rock-bottom 12.5% corporation tax. And now the weather can be added to those factors" (McDonald 2012). The climate was positioned as a positive feature of Irish space to be extracted for profit, especially post–financial crisis, which took shape in primarily public discourses. The determining factor, though, remained the generous business and tax climate of the country, and the accompanying infrastructure from which to draw, however much these discourses posed the landscape of financialization (and austerity) as natural.

This is a process that occurs elsewhere, particularly in northern Europe. Anthropologist Alix Johnson traces the variety of factors that align to pose Iceland as a "natural" home for data infrastructures (2019), and Vonderau points out this convenient terminological convergence in Sweden's north, where Facebook built a data center in the small town of Luleå, allowing state authorities to extol northern Sweden's "climate for innovation" (2019, 698). However, this sort of innovation and "clouding," as Vonderau calls it, requires a particular business and regulatory environment as much as resources and weather. As previous chapters articulate in detail, Ireland's former 12.5 percent corporation tax and a state eager to facilitate private infrastructure investment have had profoundly transformative spatial consequences. These mechanisms intensified in Ireland post–financial crisis in service of capital flows and global visibility through the country, as the state was eager to dispel concerns around economic instability and rebuild an image of robust growth potential. But their implementation also meant that companies often had free rein of the spatial and regulatory environment. This naturalized the liberties taken with the climate, and the failure to provide believable environmental credentials to back up the greenwashing. As an *Irish Times* writer found out, "'To be honest, the line on the weather is a bit of spin,' says Jason O'Conaill, international partner with Infinity Data Centres in London and formerly data centre lead with Eircom. 'The weather here is the same as most of northern Europe when it comes to computer cooling, and from an engineering point of view, well Norway, Denmark or Amsterdam would be the same really, so it's not really a strategic advantage'" (O'Dwyer 2015). This was echoed to me by the CEO at CIX, who testified that the cost savings of Ireland's cooler climate were so negligible as to not be a factor in how his company operated their data center.

Meteorological climate in the normative sense is, of course, a pressing concern for data centers. Energy for these infrastructures is increasingly sourced from the atmospheric circulations of wind and solar power. Using this transported and refined electricity, microclimates are generated within data center sheds, as the heat from the continuous power of server racks must be managed by efficient cooling, using more electricity. The whole process produces an excess of waste heat, which has been widely proposed as a circular resource for public and commercial purposes. For example, one self-described "green" data center entrepreneur proposed a €1 billion data center project on land originally zoned for data center usage in northern County Wicklow in 2008. His plan, which he had hoped to sell to a major cloud provider, was to filter excess heat from the data center through a heat pump to warm other local commercial endeavors, from greenhouses, to desalination, to a leisure center (including a biodome and indoor ski facility). This practice, to heat local homes and businesses with data center waste heat, is already in practice across the Nordic countries (see Velkova 2016), and AWS at this time was beginning to develop a partnership with South Dublin County Council to heat homes in Tallaght through the Tallaght District Heating Scheme (Moore 2020). While none of the endeavors described by the Wicklow developer were built, he said at the time that they would follow the development of his data center plan, setting in place a local growth infrastructure—"ripple effects" from the effective instrumentalization of climate.

What was remarkable about this proposal, as pitched to me, was that he insisted he had no interest in what data centers do. He merely saw servers as self-generating community heaters. Speaking over coffee at a hotel in Newtownmountkennedy, County Wicklow, he talked to me about tomato-growing greenhouses, and, mimicking holding a tomato, asked me to imagine: "These tomatoes were grown with energy *reaped from the cloud.*" His company would only be a solutions provider, utilizing the data center's continually electrified commercial activities as a *resource* for energy extraction. This literal climate extraction follows a greenwashing logic that the private sector will provide climate solutions when the state refuses, naturalizing data-driven value production as well as putting forth the private sector as the primary agent of environmental care, seeing innovation as the only way forward.

These more literal discussions of climate in relation to cooling—in tandem with the "business climate" of Ireland more broadly—form the basis of the coalescing logics that I refer to as "climate extraction," or the coordination of extractive modes of production across financial networks, supply chains, cultural conditions, and environments. In discussing the divergences between "extraction" and "extractivism," Mezzadra and Neilson distinguish that "[extractivism] provides a means of identifying the wider characteristics of economic, political, and social formations that are predicated upon an expansion and dominance of extractive activities. [Extraction] describes historical and contemporary processes of forced removal of raw materials and life forms from the earth's surface, depths, and biosphere" (2017, 185). Their argument, however, is that far from representing a new paradigm, the current financial and logistical dimensions of the global economy make it possible to "locate extractive dimensions in operations of capital that are seemingly remote from these domains [mines, plantations, etc.]" (186). The raw data of consumer activity circulating through data infrastructures, through smart technologies, cryptocurrencies, and other atmospheric modes of consumer and enterprise computing, becomes a crucial site of extraction (see Couldry and Mejias 2019; Gago and Mezzara 2017; Rossiter 2016; Srnicek 2016), where financial and logistical flows of value circulate through a mechanized array of consumer and business devices, which data centers necessarily support.[14] As Mezzadra and Neilson argue,

> The mapping of this frontier cannot be restricted to sites of literal extraction. Over recent years there has been a marked dissemination of the language of mining into other spheres of human activity. . . . In these instances, we can discern the expanded sense of extraction . . . involving not only the appropriation and expropriation of natural resources but also, and in ever more pronounced ways, cutting through patterns of human cooperation and even trespassing on the very sinews of the human body. The expanding panoply of practices in data mining is another register of this pervasive penetration of extraction across different spheres of human and economic activity. (2019, 144)

Resonant with discussions of cultural and creative labor in chapters 2 and 3, Mezzadra and Neilson argue that data-driven commerce and financialization are driven by the extraction of "skills and labor" from users and workers within these data assemblages and their modes of collection, storage, and accumulation (145). Data centers thus operate on several different levels of extraction, as the extractive economies of energy (fossil fuels), data (user info, "Bitcoin mining"), and finance overlap in one territorial base of tech capital.

Many besides Mezzadra and Neilson have recognized the conceptual overlap between "data mining" and the resource metaphors that characterize the data economy (Gago and Mezzadra 2017; Srnicek 2016; Taffel 2023), whether it is seen as new gold, currency, oil, or oxygen. However, such discussions can often dematerialize by making metaphorical literal extractive activities. What Rossiter talks about as "logistical media" describes extractive activities articulated through and across the infrastructures as well as the users of contemporary digital technology (2016). Media scholar Jennifer Gabrys, in her study of atmospheric computation, argues that the planetary imaginations of big tech companies, such as the IBM campaign for a "Smarter Planet," posits an "increasing instrumentation of the planet" (2016, 7). This logic is extractive in a way that has not been as widely theorized, in the sense that the "environment" and its "natural resources" are increasingly used in tandem with technological infrastructure in the production of more efficient circulation, without the mediating process of labor-intensive production (at least on site). Following the idea that contemporary value extraction is focused on this "noisy sphere" of circulation (Marx 1992, 279), as Mezzadra and Neilson argue, circulation and the interlocking logics and operations of "extraction, logistics, and finance" form the basis of how the state and capital partner to ensure the ongoing expansion of capitalist activities (2019). The "signal and noise" of communication in informational capitalism (Larkin 2008), movement and turbulence, becomes a site for extraction. As Rossiter notes, "the capture of value through the coordination of movement" (2016, 173) central to technologized supply chains must necessarily render global turbulence at economic and environmental scales legible for extractive activity. Data centers, as the territorial bases of digital economies, operate as transnational conduits for *flows of extractive capital*—whether tax-evading finance, the mining of user data, the shipment and burning of fossil fuels,[15] the packaging and shipping of culture, the opportunistic use of climate conditions—which takes the place of state investment in local politics and prosperity, in addition to contributing significantly to pollution and other environmental disruptions.

But in the context of "climate extraction," these operations require vast networks of cooperation, from state and private infrastructural systems to tax and planning policy, to political, social, and cultural circumstances, to environmental resources and conditions. Studies of the atmospheric circulation of data and capital demonstrate that climates are not only hot and cold, wind and rain. As Hulme argues, climates represent entangled social, cultural, and political as much as "naturally occurring" meteorological phenomena (2017). This is especially true in conceptualizations of the "Anthropocene" and "climate change" that center European civilizational agency within planetary and geological processes (see Moore 2017; Yusoff 2018). Like the universalizing images promoted by the state and capital in imagining the future, the Anthropocene glosses over historical and existing processes of colonialism, capitalism, and imperialism as an agential climatic force, as well as the resistant and unruly "worlds" that act within such planetary circulations. What would once have been thought of as the "natural environment" is not seen only as a site of extraction but rather as a site of instrumentation, of circulation, while dominant discourses produce this same environment as something under threat, to be cared for by green market initiatives and NGOs, and as "precious" as ever before. Thus, the mechanics of climate extraction colonize the entanglements of technological systems, public infrastructures, and natural environments, all loosely grouped together within "business climates" through which they operate and instrumentalize existing sociocultural environments.

Business Climates, Crisis, and Logistical Governance

Climates are both affective and representable, relational and fixed in places, influenced by capitalist activities as much as atmospheric movement. On an Ireland-specific page, Zayo, a digital infrastructure company, extolled the virtues, familiar from earlier examples, of hosting data-driven enterprises in Ireland, calling Dublin "the world's data center," claiming geopolitical proximity, partnerships, and stability; a robust and decentralized start-up and entrepreneurial culture; a pro-business environment; a "burgeoning cloud ecosystem"; a "fibre motorway" around the city; and finally "a natural climate that helps with the cooling of data centres and little threat of natural disasters, Dublin's natural economy supports this pro-business system that has been put in place"; all within the same description (Zayo 2017). The frequency and pairing of organic, ecological, and atmospheric metaphors with economic and infrastructural conditions is revealing, as though they all flowed through Dublin like the Liffey.

These associations of climate, culture, and capital in discussions of Ireland's "business climate" are especially present in corporate literature—whether "vying for some of that Irish luck" or making that luck through the turbulent Irish weather (see Watkins 2017). We can connect these discourses to the collision of cultural economies in Ireland cultivated through the Celtic Tiger and the preceding years, which offered a self-provincializing version of an idyllic rural, colonial, homogeneous, "wild" past within (and often starkly in contrast to) the heterogeneous and relatively wealthy space of contemporary Ireland, and the austerity-driven creative industries logics that thrived post–financial crisis. Like industry write-ups, Enterprise Ireland and IDA's Irish Advantage Campaign in the mid-2010s advertised the "luck of the Irish," building on this stereotype to brand Irish workforces as "innovative, flexible, and trusted," hard-working and enthusiastically ready for business (EnterpriseIrelandTV 2017). Global capital mobilizes these essentialisms in similar ways. For example, the EIB's environmental finance arm activated cultural heritage for its activities in Ireland. During austerity, Coillte was pressured to privatize forestry assets. On an Ireland-specific page, EIB described the cultural heritage of "romantic" (and modernist) Ireland, arguing that high-finance solutions to nurturing forested territory would inspire the next generation of writers, poets, and musicians (Tanklar 2017). Similar to "wild Irish" stereotypes activated by the Wild Atlantic Way tourist campaign (see chapter 3), an array of state and private actors capitalized on the Irish landscape, culture, weather, and climate conditions, extracting value from existing infrastructure, fixed capital, technical resources, and labor as well as the feelings, affects, and needs of those in a given place. As critical theorist Isabelle Stengers argues, we should follow Felix Guattari in being "hypersensitive to the danger of reterritorialization in an imaginary past" (2017, 383). This is true whether we are talking about how capital colonizes culture and heritage, or within outright nativism.

"Business climates" are promotional atmospheres generated by and about a particular place and are often culturally branded as such in response to economic and political crises to draw in capital (see Greenberg 2004). The financial crisis in Ireland was certainly one experienced and mediated culturally at the national scale (see Linehan and Crowley 2013), and ongoing crises of housing, climate, and other environmental (both built and natural) concerns were experienced and interpreted similarly. As Vonderau describes Facebook's data center in Luleå, Sweden, through David Harvey's articulation of nation-state "business climates" within the transnational and regional competition of so-called uneven development and its networks of interdependence, "What materialises in these 'national' geographies of the cloud . . . is the problematic role of the state" (Vonderau 2019, 709). While civil society in Luleå played an important part in the development of Facebook's data center, the role of the state in the cultivation of environmental conditions for extraction emphasizes the material role that governance plays, especially through place-branding, which visualizes and imagines certain territories for extraction and in doing so erases or flattens the complexities of on-the-ground politics, labor, and environments.

Post–financial crisis, the state promoted Ireland as a place to visit, make media, consume, and do business, intensifying longer histories of drawing in visitors and FDI. While "soft power" (Nye 1990), which suggests that nation-states' influence can be exerted through cultural diplomacy and long-term building of relationships across civil society and the private sector, is effective in describing how agencies such as Enterprise Ireland and the IDA promote the country, the concept takes on a certain immateriality if used without empirical qualification. Certainly, such state "branding" is effective, but also difficult to measure. Geographer Nigel Thrift has discussed the affective "atmospheres" of contemporary life as central to understanding how capital cultivates space for consumption (2008). But as in geographer Miriam Greenberg's study of New York's rebranding following a series of crises leading into the 1990s (2004), this "event-making" out of crisis is a primary tool of liberal governance, which uses the eventful relations within the public sphere to undercut and normalize the "crisis ordinary" of (particularly precarious) life under late capitalism (Berlant 2011; Povinelli 2011). Critical theorist Lauren Berlant theorized the affect of "crisis ordinariness" arising out of conditions of austerity and lack, and the ideas of these felt atmospheres of capital that solidify into an "environment," in terms of the temporality and spatiality of the "event" and its governance (2011). The 2007–8 financial crisis in Ireland, rather than changing the country's capitalist landscape, solidified its strategies and intensified the hold of financialization across various and differential environments. Here, the pro-business climate of the "crisis ordinary" masked the continued turbulent eventfulness of global capital and its circulation through Irish space— acting to acclimate conditions conducive to austerity and privatization.

As chapters 1 and 2 argue, the particularly turbulent fall in the supposedly natural global market in 2007–8 shifted blame away from the real causes and onto individuals, populations, and already struggling "peripheries," and their exploitation and suffering was designed to appear as natural, with politics necessarily articulated through financialized relations and conditions of governance, whether imposed by the state, supranational regulators, or transnational capital. Within this environment, in Athenry, an atmosphere of hope transformed into one of bitterness. However, such atmospheres also engender political responses at larger scales. For example, part of the crisis response, as imposed by the troika and its debt conditions, was the enforced privatization of public resources. The state privatized water and established a semi-state commercial utility called Irish Water in 2015 (see Bresnihan 2016), which was fought off ardently by the populace through a series of lively protests and old-school civil disobedience against proposed water charges, during which many households in the country simply refused the installation of water meters. Coillte, the semi-state forestry company which transferred their land to Apple in Athenry, was nearly compelled to privatize its forestry assets. Coillte continues to serve doubly as a resource organization and an asset manager: On a phone call with a Coillte representative, I was told that the company wants to sell more of its sites for data centers, even though it had recently sold its telecommunications wing to multinational InfraVia Capital Partners. While this might appear like two unrelated areas—forestry and communications—they sheltered under the same umbrella as the Department of Communications, Climate Action, and the Environment as of 2020. The management and protection of natural resources became asset management. They both still come into strategic cooperation as much as conflicting kinds of interests in the basic governance of how infrastructural systems operate and are developed, from the regulation of antennae to the management of natural resources.[16]

But climate finance initiatives represent even more insidious privatizations not only of traditionally public, but also of *common* resources. This extends beyond metaphors, the privatization of resources, the production of more effective business environments, and the facilitation of extractive capital: It extends into the logics that facilitate the very financialization of *air* itself, which has unexplored implications for the ideologies and politics of atmospheric computing and energy systems. Sociologist Melinda Cooper has theorized the financialization of climate through turbulence by mechanisms like climate finance and the solutions of the financial sector fixing climate change (2010). The ungovernable spaces of the "commons," long a site of valorization, become sites of future resource and energy conflict (170). As Cooper continues, "What the market in derivatives extracts from the noise and colour of the world are its event-making relations. The turbulence engendered by connectedness. Turbulence is the event emerging from an irresolvable relation between two or more 'flows' that are themselves relations" (180). The "weather" that constitutes such unmanageable turbulence, through climate finance's capture of the "noise and colour" of financial (and logistical) circulation—and perhaps the "wilder" movements surfacing through submerged histories forgotten by the state—become sites of extraction, demonstrating the ways in which it is *circulation itself* that produces value for extraction under contemporary capitalist regimes.

I choose to use the conceptual and material heuristic of "climate" over more recent discourses of "atmospheres" of capital for a number of reasons. While Cooper articulates the political economy of such relations, geographer Derek McCormack, in his studies of what he calls "atmospheric envelopment," stresses the importance of atmospheres as ways to "link the affective with the meteorological" at a "range of scales and over various time horizons," as "the significance of meteorological variations, processes, and events in this atmosphere is being ever more scrutinized as their origin is denatured, and their disruptive effects become part of the turbulent, emergent urgency of the affective life of contemporary political ecologies" (2018, 20–21). Drawing on anthropologist Tim Ingold's notion of "weather-worlds" to tease out this climate/affect relation, especially crucial because "the infrastructures upon which forms of life depend are arguably becoming more atmospheric" (McCormack 2018, 22), McCormack refers to the apparently immaterial infrastructures of computing. However, the spatialization and temporalization of "atmosphere" as generalized but felt within a given territorial formation also demonstrates the sometimes troubling ahistoricism and universalism of phenomenological, new materialist approaches to capital and affect, as suggested by geographers Monica Degen, Claire Melhuish, and Gillian Rose (2017). Branding and business interests certainly shape atmospheres; however, these are experienced and enacted differentially, and should never be taken for granted.

As Furuhata argues, climate is always subject to uneven instruments of control as it becomes more and more turbulent and subject to thermostatic intervention (2022). This emerging sphere of influence—as though it has not been one across colonial modernity—should demonstrate to us that "[t]o think about climate is therefore to think about geopolitics" (15), and thus volatile turbulence. Sociologist Martín Arboleda suggests that, in this "planetary" moment, "The very idea of the 'global' as a proverbial blue marble demarcated and measured through grids and coordinates is being gradually superseded by that of the 'planetary,' in which the earth reemerges as an unfamiliar space riddled with eerie, destructive, and menacing forces" (2020, 15). If the planetary suggests that environmental turbulence is beyond human and unmanageable, we must necessarily align conceptual tools to account for the obvious coexistence of global capital and planetary circulation, especially in the sense that we can often trace the source of the "externalities" associated with environmental degradation and the production of territorial climates within and across which capital operates. As Povinelli points out when discussing digital projects acting in response to government abandonment (2016, 164–65), even though the idea of the cloud as green and immaterial is fundamentally occlusive, this homogeneous metaphor brushes over more sinister meanings of clouds as harbingers of bad weather and nuclear explosions. The actual cloud is foggy; its externalities, smoggy. The state cannot expect capital to offset these externalities, as they try to control public narratives and visibilities while their operations exist within the clear and ongoing lineage of extractive transnational operations.

But how, then, can we reconcile the affective and experiential dimensions of our analysis in chapter 2 with those of digital design logics and the politics of data centers? One cannot discount the political feeling evinced by data centers, big tech, and the creative industries, especially in a place like Athenry, where affective politics were directly entangled with capital in a context of (real or imagined) state abandonment. The atmospheres of the "crisis ordinary" and its place-branding are affective, but, as this chapter argues, in Ireland, discourses around "climates" are perhaps better suited to reflect the intersection of aspirational and material spaces that are expressed, and the apparently immaterial externalities that represent what Berlant calls the ongoing "costs of liberal freedom" (2016, 400) exported elsewhere. Atmospheres, especially in the context of crisis, are felt, but may remain immaterial and universalized in how they articulate experience; climates, on the other hand, are observable, measurable, differential, and able to be instrumentalized across a sedimented array of processes (see Hulme 2017).[17] Such processes also account for the dissenting or "wild" circulations within a given climate, weather conditions that move against the predictable and harnessed patterns of known and accepted systems.[18]

However, as the case of Apple in Athenry can show us, these unexpected movements are not always resistant. A vast array of interests, conditions, policies, cultural histories, and ecologies present post-crisis have generated a climate from which companies are able to extract value. Whether the discursive or literal heralding of the natural climate as a source of cooling, the cultural branding of the business environment, fiber-optic cables to use information, or the activation of policy mechanisms to optimize value generation, the country has not only become increasingly dependent on extractive FDI but largely defined by it on a geopolitical level. This entangled political ecology has consequences far beyond branding and spatial development. Our ways of doing and understanding politics must fundamentally account for the emerging and differential modes of "freedom and belonging" as experienced by those places that face the entangled realities of economic and climate turbulence (Connolly 2017) in ways that are not just eventful, but that represent the endurance of certain relations and experiences. The choppy waves of global finance and oceanic circuits of trade, information, and energy reorganize and redistribute violence and instability across life and work, while a "tsunami of data" threatens to flood the world with the unmitigated growth of media infrastructures and their energy requirements (Climate Home News 2017).

These remain emergent sites of politics, where the state and capital were already gathering forces in the late 2010s. Turbulence can never be *fully* managed, and this management is designed to protect the interests of capital in the face of existing and oncoming structural instability. As consumer citizens, according to these logics, we can only batten down the hatches and ride the waves, hoping not to sink, and be ready to be run aground by the rapid recession of turbulent seas. Some would say we are lucky to have a boat in these waters at all.

We are, of course, talking about different things when we discuss data centers using a given climate—tax, atmospheric, or otherwise—to discursive and practical advantage and the increasingly abstracted worlds of climate finance. Nonetheless, data centers are part of these wider industrial and geo-economic shifts and sovereign disputes emerging in the face of a climate-changed planet. All these factors make them strategic nodes for both capital and its discontents, and sites of securitization and conflict. Green development is a firmly entrenched strategy of capitalist valorization, especially through how it utilizes places for development *because of* given climates or weather patterns. This is a direct financialization and logistical management of climate, as secure facilities like data centers capitalize on "cool climates" while industry execs admit that this is mere corporate spin to take advantage of friendly tax climates, an environmental twist on an issue of geopolitical competition. Data sovereignty and the increasing instrumentation of the planet have global consequences, but we can find within localities of friction and struggle the conditions through which these projects are either strengthened or disrupted, especially in and through environmental politics. But just as "infrastructure makes worlds," so "logistics governs them" (Rossiter 2016, 5), as the private administration of goods, data, and people comes to operate this peculiar strand of nondemocratic politics, presenting the new local and transnational frontiers of circulation struggles.

But it remains true that the climate, the planetary circulation of air, gases, weather, and water, is fundamentally ungovernable, necessarily tied up within transnational geopolitical and economic negotiations and regulations, fought for bitterly on the ground by environmental activists against agents of state and corporate violence, but also turbulent in ways that evade harnessing and enclosure. As much as the natural climate becomes a source for speculation and control, as it always has been, its status as a site of *future* conflict remains an abstraction (although not for long). Just as residents in Athenry feared current dissent would dissuade future investment, as though its flows are as fickle as water, the future is not already here, as the state and their corporate partners might try to project.

Berlant, in the aptly titled essay "Infrastructures for Troubling Times," argues that "institutional failure leading to infrastructural collapse, from bridges to systems to fantasy . . . leads to a dynamic way to disturb the old logics, or analogics, that have institutionalized images of shared life" (2016, 403). Thus, political and cultural imaginaries and the commonsense logics of public and private in the case of the peculiar and intensive infrastructures of data centers (and similar data infrastructures) perhaps need a new set of conceptual tools, or they risk succumbing to "regulatory hangover" (Holt and Vonderau 2015) and a return to institutions that are geared irreversibly toward perpetuating the system they would be tasked with regulating. One sees these responses take shape in widespread support for multinational data centers, or consent for their role in shaping our futures, as somehow "green" versions of existing extractive activities. Global capitalist development discourses and their geopolitical implications, especially surrounding Ireland, are increasingly placing real and cultural capital on the "climate" and its extremes (as well as its everydayness). The logistical operation of corporate data infrastructure within a wispy, vaporous cloud pushes a kind of sovereignty that keeps data (and its capital architecture) untouchable by alternative political futures, by the strategic capture and mobilization of the global commons.

Conclusion: Weather Patterns

As this chapter illustrates, financialized industrial modes of flexibility, optimization, efficiency, and automation dovetail with property development logics of media urbanization and rural development, the productive rationalities of media production, and the logistical technologies of data centers. In each case, the state plays a crucial role in promoting and facilitating the various ways in which these are implemented and allowed to function across the space of contemporary, post–financial crisis Ireland. In data centers, we see the ideal location within which people—political subjects living in the country, diversely in support of, opposition to, or unaware of these areas of the economy—can be dispensed with almost entirely, and a self-perpetuating technology of accumulation appears to be operative. However, as is true with each of the other cases, the actual conditions within which people, communities, and subjects relate to the state, capital, and the environment are far messier, far more unruly, than the world-eating mechanisms big tech would want. Places respond to turbulence in their own way, in unpredictable ways. Whether local communities living and expressing themselves in tension with global corporate strategies, lone objectors standing up to the state and global capital, data center developers accidentally providing too much information, or cheeky security guards briefly lapsing in their role as protector of these sheds of information, data centers, their social and managerial realities, their physical footprints collide and produce unpredictable encounters, outcomes, and relations.

As someone who has continued to do research on the Irish data center landscape amid a continuously expanding and innovating set of industrial conditions, it is hard to draw a line to stop analysis, especially as their reach and influence in Irish (and global) infrastructural politics has continued to grow. I did not predict, for example, just how much artificial intelligence (AI) would come to shape data center development and expansion in 2020, not least because of that industry's lack of penetration in the Irish landscape until much more recently. However, in the period traced in this chapter, toward the end of a long decade of austerity that gave way to a "recovery" and now a COVID-19– and AI-changed world, we can see some of the methods by which big-tech capital represented by data centers came to be naturalized as an atmospheric form of circulation within Ireland's infrastructural landscape, as well as agents of environmental care and influence over green policy. Data centers were key sites at which the rationalities of finance capital and the infrastructural build-out to support new media technologies coalesced in the climatic conditions of financial crisis Ireland. This was achieved by extracting *for* and *from* data circulation through a built and "natural" environment already coded with colonial residues and economic development logics of hyperbolically green wildness and rurality ripe for cultural, touristic, industrial, and financial exploitation, all at the same time.

The infrastructural and industrial ruins of former periods transition into new modes of productivity, with as-yet-unseen consequences once these private infrastructures come to decay, and society's speculative gambles with transnational capital as purveyors of care lead to new asset bubbles, and new bursts. New crises, turbulences, and regulatory changes leave places like Ireland, intricately tied into interdependence by an FDI-driven economy, more precarious. While the servers and computing stacks within a data center are under near-constant renewal, with technologies only expected to last for a few years at a time (Ward and Carcone 2015), the buildings themselves are subject to their own turnover, repair, decay, and disappearance, becoming "cloud ruins" to supplement the financial ruins in the built environment (Brodie and Velkova 2021). With global tax reform finally on the docket,[19] amid changing business climates and regulatory approaches to tech, and considering the constantly evolving innovations in digital storage and efficiencies,[20] what will happen to these buildings once they become redundant or obsolete? Most importantly for Ireland, and for those gambling on the future social and environmental care offered by tech companies and their private infrastructure, what will happen if the data center bubble, like the property market before it, also bursts, and the investment recedes?

As this study of Ireland's climate and advanced extractive processes can tell us, in conversation with local frictions, political climate can change, and is always changing, based on currents and patterns happening above, on, and below the ground. As the 2021 EirGrid curb on data centers in Dublin demonstrates, the conditions that transnational capital encounters in Ireland are not nearly as fixed as the state promotes. There are different weather patterns to follow, and rain and wind can always clear at a moment's notice. But it is a matter of changing political behaviors and beliefs that will ensure that these climatic conditions are or become stable and represent a healthy environment for people to live and build collective power in.

Such changes must also come with a recognition that a particular climate is only one among many, which must coexist with others to ensure a healthy and balanced planetary system. Interconnectedness at a variety of scales needs to be constantly in our minds, especially in light of increasing turbulence. If one climate remains geared toward extraction, then others are not safe from the externalities of these activities. As Povinelli suggests (2016), fog becomes smog in the externalities of the cloud, and industrial runoff and toxicity are still felt most violently among the marginalized and in the Global South. Politics in Ireland must perform with an understanding of the country's place as a perpetrator within global systems of inequality, within which this FDI strategy not only enacts uneven impacts across Ireland's urban and rural environments, but beyond its borders through global systems of finance and extraction. Understanding how multiple, overlapping, and intricately entangled territorial climates across the world are both anthropogenic and not at the same time, human and nonhuman, controlled and uncontrollable, tamed and wild, in the sense of the constant entanglements of these domains of activity, allows us to see the simultaneous and multiple scales at which we must focus our conceptual and political energies.

The fact is that the sphere of the climate that organizations—whether the state or capital—are attempting to condition, manage, and control tells us much of what we need to know about the stakes of the naturalization of global circulatory systems. There is nothing natural about the machinations of transnational capital, but these operations occur within planetary systems nonetheless, as much as across borders and global contexts. Thus, there is always turbulence and disruptive movements when capital hits the ground and tries to operate; planetary systems continue to thwart attempts at control and management. This should encourage us to deploy our politics toward disrupting global systems—and their imagined futures—and protecting planetary ones at all costs, even if the planet seems increasingly turbulent and frightening (see Arboleda 2020). Seeking stability should be no excuse for complacency, as we know how quickly the tides can turn, and the planet thwarts such inaction. An alternative politics of climate is urgent, and the state and capital are already ahead of the curve, focused intently as they are on controlling—and harnessing—this sphere of activity.

Conclusion

Calmer Seas

Let us conclude by returning to the central metaphor of *tides* from the opening pages of the introduction. As the high waters of finance capital receded, years of austerity left the salt to harden across the landscape, revealing the ruined, corroded structures that had been left behind by the waters. These social and material formations had been laid bare by recession and left to decay by the bitterness and affective politics of austerity. Young people fled; cultural and public funding was cut; the commodification of culture and landscape intensified; labor became more precarious; inequality grew; and all the while, global capital continued to function in this landscape, even as many suffered within earshot of its machinic hums operating behind enclosures, nourished by subterranean currents that others could not access, channeled in through proprietary pipes.

However, before most had even figured out new ways to live in this barren landscape and its novel arrangements, the formations that emerged in the mud or crystallized in the salt flats had scarcely dried before a new, soft tide of foreign capital began to tentatively creep through the country, fully immersing certain areas once again, dissolving the logics that had coagulated and dispersing them into the mucky waters. The operations of foreign capital pulsing through Ireland's economy reveal a more muddled and toxic version of what had caused the destruction in the first place, and the tentacles of multinational companies lurked in the new and murky pools. What the project that makes up this book did, perhaps, throughout nearly real-time research, was keep track of how these formations dissolved into these turbulent waters, and how they affected the emergent ecosystems—not only in continuity with older forms of rule, extraction, and exploitation, but also in new and poisonous ways, and before they began to disappear once again under the auspices of prosperity, or worse, emerging crises.

I now live in Ireland, and these cyclical tides are perhaps more daily experienced as trudging through permanently muddy and under-functioning infrastructure and services. Even with a relatively cushy job at a good research university, my everyday experience of austerity and crisis affects all prospects of life and work: My commute is two hours each way, I rely on a private regional bus service that is terminally stuck in traffic and goes by a locals-only schedule, my contract isn't permanent, research funding is increasingly prescriptive and based on commercial imperatives, and, most prominently, the cost of living makes a tightened belt a fact of life for anyone not pulling in an extraordinary wage. In late 2023, on the bus to work in south Dublin, an ordinary morning politics email column from the *Irish Times* read thus:

> "We have been saying that a moment of change will come," Paschal Donohoe [minister for public expenditure, national development plan delivery and reform] gravely intoned, informing reporters yesterday that the corporation tax take in September slipped to €1.8 billion in September, €300 million less than during the same month last year. Overall, corporation tax is down 23 per cent compared to the same period in 2022. The dip now looks likely to become a trend. . . . The Minister for Public Expenditure is correct—he, Minister for Finance Michael McGrath (who said yesterday corporation tax income was still growing, but at a "dramatically" slower pace) and their officials must be hoarse with all the warnings they have issued about the volatile nature of corporation tax receipts as corporate profits flooded the exchequer. Now, on the eve of the coalition's most delicately balanced budget, these warnings are being borne out. . . . They cannot plead poverty, but for the time being, the mood music supports their firm stance for prudence. (Horgan-Jones 2023)

An attentive reader will notice the naturalizing metaphors and crisis language permeating here. The barely coded warning from political leaders and decision-makers is that the Irish population would suffer as a whole—at the hands of a budget that many Irish residents are desperately counting on to support them through public services and expenditure—if corporation tax receded, meaning the "mood" supported a more austere budget to tighten belts in advance. The recognized volatility of corporate profits combined with the doomed inevitability of crisis and austerity: not only political economic theories informing academic research and state decision-making, but these were also everyday experiences in post–financial crisis Ireland.

This book has traced some of the ways that this environment took shape in and through media infrastructure. But when it comes down to it, this remains a fact of life for millions of people here. We are beholden to corporate profits, but the benefits are somehow always withheld from most—they are almost here, just a few million euros more, and we'll finally feel them being more widely distributed. In the meantime, rents rise, house prices climb, cost of living soars, public welfare support diminishes, and wages stagnate. To be frank, Ireland is a place of almost cartoonishly naturalized austerity. Turbulence is experienced as an ongoing condition of dysfunction, as its enduring consequence has been not truly "privatization," but the evacuation of infrastructural and institutional capacity for managing anything other than the growth and facilitation of corporate profits. Any rising tide only lifts yachts for those in the Docks or Silicon Valley.

As I have suggested throughout, paraphrasing the anthropologist Aihwa Ong (2006), the interfaces between the state and capital affect daily life *directly* in an era of transnational capital, *especially* in conditions of precarity and crisis, far more than might appear on the surface. Drawing once again from another anthropologist, Anna Tsing, perhaps the "precarity and indeterminacy" that are posited by normative liberal politics as marginal to systems of capital and rule are in fact central to how global economies function in the first place (2015, 20), across dispersed territories of operation. The structural role of such instability is particularly urgent when it comes to climate change and the environment. As opened by the final chapter, the literal dimensions of the tides and toxicities conjured by this book's central metaphors are becoming more and more pressing. In an island country, these tides are undoubtedly more urgent and more extreme than in many places. This is why data centers, where this book ends, are an instructive media infrastructure for understanding the emerging contours of a climate-changed landscape in an era of seemingly permanent, or at least naturalized, austerity. I have used data centers as an opportunity to bring together various strands of investigation and point them in directions for future study, theorizing and laying the groundwork for how we can better view the sites at which emergent forms of rule are coalescing, as well as places where unruly politics might represent frontiers for future struggle. Climate change, and the chaos that it is already bringing, is being inextricably tied into novel infrastructural formations with the "twin transitions" of digitalization and decarbonization, and being amplified by the extraordinary infrastructure boom to support AI growth. This book thus brings us *nearly* to the present, analyzing a place where these politics have been playing out in a concentrated environment of austerity.

In the intervening years since the conclusion of this study, the degree to which the financial crisis laid the conditions for an increasing experimentation with private, extractive, multinational media infrastructures has become clearer. Projects like the National Broadband Plan have been rolled out with the expected public–private partnership and incentives to varying degrees of success, while the integration of multinational tech capital and the ongoing abandonment of ordinary Irish residents to underdeveloped state infrastructures in the face of mounting crises represents the ongoing radical divides between who benefits from "green" development programs and who does not. The conditions traced in this book, which demonstrate in some cases early and speculative attempts to strip public assets and wealth and direct it toward the private sector, have only intensified. Ireland, while for most of modern history a place of confounding contrasts and apparently irresolvable contradictions, is an exceptionally unique place to live. As a small country emerging from a relatively recent period of postcolonial underdevelopment and currently seeing obscene wealth circulated through it, the comparative unavailability of public services and infrastructure should tell us everything we need to know about the "public" benefits and investments of FDI-driven development. Youth especially continue to flee for less turbulent shores, and you cannot blame them. Students in my classes at UCD in south Dublin frequently commute from places farther than my own, already serious, commute from Dundalk—and most often from their parents' houses, as accommodation in Dublin is not only unaffordable, but unavailable, full stop, and the cost of living continues to soar.

Under these circumstances, the pervasive feeling that another crisis is on the way should probably be recognized as the internalization of the "crisis ordinary" (Berlant 2011) that has never subsided. Its processes and logics have become so naturalized as to be the only way to do government, policy, and planning. It cannot be overemphasized how turbulence, how much *whiplash*, has been experienced in Ireland since the 1990s—from exceptional, unprecedented growth and flooding of wealth to a profound and incapacitating crash, these two economic conditions have shaped life for the foreseeable future in Ireland. The conditions that drove the Tiger in the first place remain, with all its inequality and FDI-driven growth, as do the structures of austerity and privatization, with all the gutting of public services and redirection into private profits. In this environment, multinational corporations have taken even greater hold over state policy and the viability of life and livelihoods in the country. Objection to these systems remains ostracized, or at the very least subdued by center-right government policy and media discourse. Social, political, economic, and environmental injustice of each condition has been continually compounded by the other, in a way that feels cruel, perverse, and never-ending.

But as I hope this book has demonstrated, while these contemporary pathologies have emerged in a particular environment of post-crisis austerity, they are historically embedded in longer processes of economic development, from the developmental to the post-developmental state in Ireland (O'Hearn 2000). The Celtic Tiger, a post-developmental phenomenon driven by exceptional deregulation and incentivization of FDI and property development, was itself the result of retooled liberalization that had been under experimentation since the late 1950s. These experimental conditions, within which FDI played a central part, tied the ideologies and infrastructures of Irish "prosperity" to the turbulence of a global market. In doing so, they generated the material conditions within which contemporary transitions are unfolding across the media and environmental relationships of a climate-changed planet. In these logics, as this book has argued, the differential and extractive management of the Irish environment—in particular its space, infrastructure, resources, climate, and labor—has led to a situation where the state and multinational capital together hold the reins of Ireland's environmental futures (see Bresnihan and Brodie 2025). This should be enough to worry us, although I hope this book has raised other, more serious questions about the sustainability of this post-developmental model.

What is important at this point to emphasize for Ireland's future is that while the *natural environment* is by necessity in the crosshairs of the presiding state and cultural rationalities of finance and logistics through sustainability initiatives and climate finance, the series of *naturalizations* processed through this book are only the beginning, as we have yet to fully grasp the implications of climate change for future spatial governance in Ireland. But the gears are turning. In the Project Ireland 2040 policy framework, there is an extensive focus on sustainability and branding Ireland as "green" framed through the concept of "natural capital," which acclimatizes the optics of capital toward the various ecologies represented within Ireland, framed through pride of place and commodified versions of national landscape and culture: "Ireland's environment and its diverse landscapes form part of our 'green' persona and we have much to be proud of" (Government of Ireland 2018b, 116). Like labor and the climate, the country's environments are treated as resources. This extends beyond climate finance and renewable energy infrastructures, and into more traditional media industrial considerations, as Project Ireland 2040 frames it in terms of preparing for and managing climate change: "Our coastal areas are also a key driver for Ireland's tourism sector, which the successful branding of the 'Wild Atlantic Way' and internationally recognised location shoots for the film industry have highlighted in recent years. Ireland's coastline is a remarkable but fragile *resource* that needs to be managed carefully" (103, my emphasis). And while this, as Curtin argues (2016), might in a more just world lead to forms of resource stewardship, of repair and of care (Mattern 2018), the overrepresentation of FDI across the rest of the plan suggests that the reality will be more intensified forms of extraction. The climate and the culture are predictably naturalized for capital.

As data centers and the spatial governance of the tech industry demonstrate, these extractive mechanisms are placed at the center of policy, and the private management of their turbulences means that the public is left behind. The wealthy and powerful will ride out coming storms in controlled environments, protected and secure and climate-controlled via ongoing energy extraction, while those making do on the outside are left to drown in the churning waters or burn on the arid landscape. It is already happening, and arguably has happened for most of modern history. What the latter portion of this book presents is an analysis of the interconnected series of naturalized policies and conditions that generate the global economic climate and reproduce its existing mechanisms.

However, as I hope I have also demonstrated, there are always, perhaps "wild," resistances that evade the representational and controlled systems of contemporary capital and governance, usually in highly localized formations or in unexpected resonance across territories. Throughout, I have gestured toward alternative and unexpected formations, despite perhaps foregrounding the dominant role of global capital in bringing turbulence.

As the cases studied throughout can prove, capital, labor, communities, and the environment always exist in a variety of *fields of relations* to systems of power. Anthropologist Llerena Guiu Searle tells us that the stories that capital relates are important to understand how it creates the conditions for its own systematic continuation (2016). Postcolonial theorist Dipesh Chakrabarty famously recounted the "two histories of capital," which splits historical view of the capitalist development to understand the structural role of colonization and subalternity within its operations (2000). We understand now how these histories are living, how capital continues to conjure new formations out of turbulent tides, offering ethereal stability for the subjugated while sowing discord and violence. Industrial modernity, promised by liberalization and its globalizing infrastructures, was merely one of many formations across the world replaced and supplemented by neoliberalism, financialization, and supply chain capitalism. As chapter 1 contends, there is no such thing as a logical progression from premodernity to postmodernity, underdeveloped to post-developmental, precolony to postcolony, precapitalist to postcapitalist. However, as Tsing argues, "This is a story we need to know. Industrial transformation turned out to be a bubble of promise followed by lost livelihoods and damaged landscapes. And yet: such documents are not enough. If we end the story with decay, we abandon all hope—or turn our attention to other sites of promise and ruin, promise and ruin" (2015, 18). As Karl Marx identified in his early theories of circulation, capitalism is structured by continuously evolving phases of production and accumulation. Cycles of promise and ruin exist in a corresponding field of pasts and futures, and infrastructure is uniquely a site through which to understand the simultaneous existence of promise and forthcoming or existing ruin, especially in relation to how the institutions of global capital continue to spin chimeric yarns about prosperity and stability. What looking around us and surveying the ruins already here can tell us is how we can begin to imagine autonomously from these institutional frameworks, which bind current and ongoing ways of living to colonized pasts and capitalized futures, both of which enact structural violence on those promised and aspiring to a different world.

While this book begins with a description of post–financial crisis politics, after chapter 1 and the discussions of "ghost estates" in chapter 2, it focuses on what we might retrospectively call the "recovery era." The "Celtic Phoenix," rising on its wings of foreign capital from the ashes of the Tiger's spectacular flameout, began its life cycle anew. I reject this imagery because of its ascription of a symbol associated with an ascending and unified republican nationalism to what is ultimately a zombie-like imperial system of value and rule. Research has punctured this numbers-driven salvation story as, at the very least, "clearly misleading" (Honohan 2021), and at worst "leprechaun economics" (see Regan and Brazys 2018), completely ignoring the mounting social, political, economic, and environmental contradictions entailed by the numbers heralding "recovery." However, perhaps what even those who buy into the Phoenix must tacitly admit is that the Tiger must have itself been a Phoenix that had combusted previously: perhaps the aborted dream of a sovereign socialist republic itself, which has been supplanted by a crass form of what Patrick Bresnihan and I refer to as "FDI nationalism" in the mobilization of this development model and its infrastructural logics (2025). Whatever the case, even proponents of the Phoenix must realize that this bird will ultimately suffer the same fate, burning up as it reaches its apex, leaving ash and ruin in its wake. The promise of prosperity, tied to that of infrastructure, is seductive and chimeric amid capital's global turbulence, and its material consequences are a map of affective attachments to its fickle circulations.

These narratives and their attachments are hard to short-circuit, and I have tried throughout this book to analyze and in doing so disrupt the teleological stories they tell. However, we may once again find Tsing's methodologies inspiring here, in terms of looking at the more unexpected things arising out of such formations: "To live with precarity requires more than railing at those who put us here (although that seems useful too, and I'm not against it). We might look around to notice this strange new world, and we might stretch our imaginations to grasp its contours . . . to explore the ruin that has become our collective home" (2015, 3). Ireland's ruined landscapes post–financial crisis were revealed to be exceptionally vital, at the cusp of new uses by both capital and the communities piecing together a living in the wake of finance's toxic tides. While the logistical imaginary of the state seems to be winning out, I hope that I have pointed to places where people have negotiated ways of living, blocking the drains themselves and figuring out ways to live, work, and struggle within and against these undernourished and undernourishing ruins of capital. For Tsing, the landscapes of matsutake mushrooms offer an exceptional place to view how nature and culture exist in the ruins of capital—whether in the stripped industrial conifer plantations where they grow, or the markets and social relations that they otherwise produce. This book's focus on *systems* themselves means that the interrelations between capital and nature are hard to parse, with the very climate of the country part of an emerging politics of rule entangled with the logics of financial and logistical capitalism. While these are ruinous circulations leaving spatial wreckage behind, the "recovery" was a time where the flows were crafting new and perhaps even more precarious structures once again.

Thus, the continuing grasp of these processes in Ireland through the dominance of the FDI model means that the wider built assemblages—the infrastructures of capital's circulation—are where we need to direct our continuous attention. This is true for unraveling how they govern, for identifying how they deal with friction, as well as for attending to the potentialities opened at these points of tension and conflict. What if we can come together to redirect, to build a levee against, to plug the drains of the unnatural tides of global capital, while building new systems of collective care among one another?

Examples throughout this book already point to sites of emergent media politics: media workers and practitioners continuing to find resourceful ways to subvert the property markets in Dublin with alternative forms of spatial practice; artists working together to form social infrastructures and community networks of support; residents adjacent to development projects leveraging their needs across existing relationships to culture and the landscape. In the midst of ongoing and intensifying exploitation and extractive activities, workers collectivize and communities come together (and come apart). People continue to relate politically to the projects of transnational capital, making claims and working within and outside of available systems. Living together in these ruins will require forms of cooperation that are already being performed and will have to be activated based on local cultural and material conditions, leaving space open for collaborating and performing solidarity with aligned struggles, however difficult, ambivalent, and ad hoc such connections may be.

The presentness of the cases throughout this book created methodological, practical, and conceptual difficulties, especially in a context where these processes endure, transform, and intensify to this day. How does one study such an evolving and unstable set of circumstances? How do you even know where to look? Ireland offered an exceptionally vital place through which to understand these phenomena and pick out points of struggle, find things that look and feel important, even as their conditions of emergence mean they may disappear or be foreclosed at any moment. Nonetheless, many of the cases and aspirational projects traced in this book are still ongoing, or perhaps even changed, out of sight, since the research project has been "completed." Such is the nature of much media research: In attempting to trace threads as they happened, these threads continued weaving different stories in directions and timelines far beyond the scope of this book.

As such, I have pinpointed key moments and sites of encounter at which these projects were at certain thresholds of emergence and tried to find the politics around the edges that could tell us more about what exactly they mean. The state and its corporate partners try to manage the turbulence from centers of operations, but it is at these extremes, at the edges, where the biggest swings are felt, even if in a delayed fashion. Related to such delays, caused as much by financial turbulence as by the Irish state's messy planning procedures, many of my cases have more stories to tell: NAMA planned to repay all debts by the end of 2020 (National Asset Management Agency n.d.), raising a question as to the fate of the organization and its sunken assets once its exceptional remit was "completed"; investors in Dublin Bay Studios, described in chapter 2, who advocated strongly for a site in the Docklands, announced that they secured space next to Google in Grange Castle Business Park South to build the "Grange Castle Media Park" (Hamilton 2020); the *Star Wars* screen tourism operators in west Kerry, from chapter 3, were tasked with enduring the COVID-19 shutdowns of tourist economies, meaning the lack of state funding as a "creative industry" became even more pronounced; and after Apple shopped their abandoned site in Athenry to foreign investors as a "ready-to-go data centre development site" for an undisclosed price (Daly 2019), they reopened their planning permission to keep the door open for a future sale to another developer. The Athenry community continued to strive for economic inclusion, which may ultimately mean becoming an energy frontier for the energy-starved data center ring expanding outward from Dublin.

Let this, then, serve as an acknowledgment that *Wild Tides* presents an approach to the entangled phenomena of media infrastructural relationships more than a holistic analysis, a series of coalescing cases and sites rather than a map of a total system. I have tried to avoid reproducing the planetary "mastery" of space implicit within the projects of global capital, or even the smaller national-scale totalities built within something like Project Ireland 2040. The spaces I conjure here in these pages are incomplete, as is any recorded memory of events. But as I suggest in the introduction, perhaps in this environment of instability, we should continue to seek and nurture that which evades representation, what is revealed only in the leaky thresholds between everyday life and the systems within which we try to get by, in the histories that emerge suddenly from the landscape, breaking representations of dominant systems. After all, we are often drawn to these moments, to the unexpected, that fundamentally short-circuit how we see the world and politics. In moving between such moments across this book, I have, of course, missed some monoliths—like television and radio, two enormously influential and still robust public media infrastructures in Ireland—as well as infinite other histories, currents, and ecologies submerged in the murky waters. So much methodology reads like a justification for bad methodology, and I am very conscious of reproducing this trope here. However, any book only has room for so much, and each of these absences, far from any form of intentional erasure, can hopefully open capacious avenues for study, both for myself and for future readers.

Nonetheless, two major events in early 2020 offered an undeniable end point for this book, without which the continuous strategies of the post–financial crisis economic "recovery" would still be forging ahead under Fine Gael's ruling government. The first was the leftward electoral surge in the February 2020 general election, which saw left-leaning Republican party Sinn Féin evenly split votes with center-right Fine Gael and Fianna Fáil for the first time in modern history. Far from a purely Sinn Féin electoral victory, what this election represented was a broad coalition on the left across parties ("vote left, transfer left" was the popular slogan on social media and in community organizing) and an encouraging closure of the disconnect between an increasingly and vocally progressive Irish populace and the social and fiscal conservatism of their elected officials. With any luck, this signaled the beginning of the end of these two center-right parties' stranglehold on mainstream Irish politics, although the Frankenstein's monster Fine Gael-Fianna Fáil-Green coalition government proved how unwilling those in power are to hand the reins over to the country's populace. (And, it should be noted, the 2024 elections, which once again saw Fine Gael and Fianna Fáil sneak into government, in spite of further Sinn Féin gains, were a discouraging sign.)

The second event was the economic recession triggered by the COVID-19 pandemic. This economic crisis was very unlike that of 2007–8, although it laid bare, once again, the cruelty of economic and political systems both in Ireland and across the world, especially in the absence of social safety nets, housing guarantees, and health care for all people. IDA Ireland had to steady fears that FDI would shrink catastrophically in the coming years, as they could no longer host potential investors and show them sites for development with travel restrictions and liabilities (Goodbody 2020). Once again, the precarity of total reliance on FDI—which here also coalesces with the revenue of international tourism—was particularly striking. However, although many were quick to recognize this difference as one that transcends economics, the fact is that it was a crisis whose conditions were made possible by the environmental destruction of the "Capitalocene" (Haraway 2015; Moore 2017), whether in terms of the encroachment of human activity on animal habitats or the carbon-heavy mobility of wealthy people across the world. Whatever happens, 2020 was and will continue to be a pivotal year in world history and for the island of Ireland, offering an ellipsis, rather than a period, by which to fade out this book. I hope that the cases traced throughout offer apertures to cultural and political futures from which we can learn for current and oncoming crises, in how academics, policymakers, and activists can unmask and unravel the actual forces of privatized production and circulation, and the power behind them, which hold far too much sway over our cultural and political lives. The logics unearthed across this book demonstrate the extent to which the state and its corporate partners are willing to condition infrastructure, labor, and the environment for extraction at the expense of more just ways of living, working, and relating to our environment. Working conceptually and practically across these fields, then, is an essential exercise for not only trying to navigate increasingly choppy waters, but for looking beyond managing ongoing turbulence and steering ships toward calmer seas.

Coda: No Trespassing

We will end by returning to the woods in Athenry. When entering Derrydonnell in 2019, a sign referenced the Occupiers Liability Act of 1995, reading, in all caps: "If you pass beyond this point you are on a premises. Please take note that the occupier of these premises given the nature, character and activities of these premises hereby, in accordance with section 5(2) of the Occupiers Liability Act 1995 excludes the duty of care towards visitors under section 3 of the act. Warning: unauthorised entry is prohibited. Private property. No trespassing" (see figure C.1). The occupier (Apple) grants no permission to set foot on the land and takes no responsibility for your care upon trespassing—on *their* land. As if there was any doubt about the "duty of care" that Apple would have felt toward the people of Athenry, the law states outright, upon entry into *their* territory, that there is no trespassing, this land is no longer public, and citizens not only lose the responsibility of care upon entering, but are outright forbidden from using the space.

Seán, and the Athenry for Apple group, had consented to this occupation, in a way, by supporting Apple's project. The state had paved the road

C.1 Notice at the entrance to the planned Apple data center site in Athenry, summer 2018. Photo by Patrick Brodie.

for it. But Seán had a far more unexpected relationship to this land than one of pure instrumentalism. We walked straight past the "no trespassing" sign as though it was not there. He felt *some* manner of ownership, of right, of *belonging*, that evaded the property laws and logics of the state. Unlike state-centric notions of citizenship and belonging that tend to characterize political theory, Seán's sense of belonging was tied to the land itself, his proximity to it, his family history within it. We were not trespassing that day. Even if Apple had built a data center there, Seán still would have belonged. Different political relations to the land, and thus alternative futures for how it will be used, hide in plain sight.

This sense of strong political ties to land, especially in rural and agrarian contexts, has a long history in Ireland, where land politics have motivated revolutionary struggles for centuries.[1] How else would a populace mobilize when a colonial power constantly uproots and expropriates from the territory on which you work, engage in culture, make a life? These politics have continuities in the contemporary Irish state, but also concrete ruptures and emergent politics. The government, and the nation-state over which it presides, does not determine belonging within the various spaces across the country mapped throughout this book. Multinational tech companies, while given the keys to the kingdom, likewise should not be determining the future spatial politics of the country. Those with material stakes in the land, whether those with a long history or those who have just arrived, deserve to self-determine such spaces and make their lives there. This should begin with needs defined through democratic and community-oriented modes of spatial practice, centered on social and environmental goods and flourishing—fulfilling the revolutionary promises of anti-colonial liberation that once motivated land politics here, rather than the regressive and limited promises of the state, capital, and worse, the nativists who misappropriate these histories toward exclusionary ends.

After all, more historical sites of radical contention continue to see active political mobilization. We can look, for example, at the Shannon Airport, run by the remnants of Shannon Development Corporation, which governed the Shannon Free Zone and Shannon town, where Ireland's purportedly "neutral" state allows US military planes to refuel on trips to undeclared wars in the Middle East—including, most recently, as an allied power with Israel. This unpopular and unjust arrangement has remained amid the genocide in Gaza and historical support in Ireland for Palestinian liberation. In 2012, Margaretta D'Arcy, an octogenarian actor and activist, broke through the airport's perimeter fence and shut down international flights in protest of these operations. Margaretta did jail time for this action. Affiliated groups have performed similar acts of trespassing and civil disobedience, including partially dismantling an airplane on the tarmac. The arrest of the "Galway 3" for trespassing at the airport in 2024 has been the most recent of these clashes between the Irish left and the Shannon arrangement. Such a nonrecognition of sovereignty motivates most political actions in the age of circulatory capitalism, as I have previously identified in terms of "circulation struggles" (Clover 2016), including pipeline protests by indigenous groups in North America and logistical struggles at ports and on highways by transport workers, including workers organizing to block weapons being shipped to Israel. These circulations do not rule us, even if they increasingly permeate the forms of rule that disproportionately subject some over others to state violence. The nonrecognition of such imperialist sovereignty takes different shapes and scales. Like the residents of Athenry ignoring the sign marking the Derrydonnell land as "occupied," activists in Dublin break into vacant (and occupied) buildings to build alternative housing politics, farmers park their tractors in the middle of Dublin traffic to contest unpopular regulations, and D'Arcy and her comrades in the Galway 3 cut the runway fence to oppose the US military. While completely different in political orientation and horizon of struggle, all entail acts of radical nonrecognition, of refusal, of blockage. And this should open apertures to a different sort of territorial politics—a crude diagram of the future alignments that would be required toward more just and democratic politics of land and economic development, even if true emancipation remains only a flicker on the horizon.

What these anecdotes tell us is something that we already know: Extractive enclosures created by transnational corporate power are not natural, nor are the logistical and financial flows of global capital. But those made by states are also not natural, and their systems of rule not as powerful and all-encompassing as our theories of politics often suggest. Spaces are always full of alternative and turbulent possibilities. The sense of belonging policed and administered by the state would not necessarily have compelled Athenry residents to follow the laws set down on this piece of property. Residents would have kept using that land until Apple put up a fence. But as D'Arcy's and others' trips across the runway in Shannon can show us, fences are fallible. They can always be cut, climbed, and removed, opening up new ways into such enclosures.

While Apple is long gone, its tide receded, public discourses and residents of Athenry still cast the space as somehow foreclosed, its use tied up elsewhere, its structures crystallized by the rescinded promise of a coming wave of capital. However, there is a different way to see this space. Derry-donnell Woods are growing back, wild and out of sight, except for those who continue to walk and live by them. These ongoing and future growths sustain in different ways, in ways difficult to render profitable. We can only hope that such growth outside of capital can be nurtured as we figure out how to live better beyond its ruins, before the tides carry even them away, and us with them.

ACKNOWLEDGMENTS

Like some of the development projects traced in these pages, this book has been under construction since 2017. While writing it, it has really become a historical work—many of the basic factors traced in this book are now almost twenty years old, providing an almost longitudinal analysis of both cultural and spatial phenomena, as well as different ways of thinking about the world economy and its overlapping crises. Those influencing and inspiring my thinking have changed over time, and this book is a record of those debts to the scholars, friends, and comrades who have tirelessly worked to understand and dismantle the durable structures of capital that continue to erode the infrastructures and possibilities of livable and meaningful futures in Ireland and anywhere else. I want to thus acknowledge and remember those who continue to face the buffeting winds and waves of global capital—or, shall we call it what it is, imperial violence—and still manage to provide us with pathways to a better world, from Ireland to Palestine to Turtle Island/North America, to all corners of the planet.

There are, of course, those people and organizations who are directly and indirectly in the DNA of this book, and whose guidance and inspiration cannot be captured in any such planetary gesture. I had the opportunity to come to Ireland several times over the course of the research for this book, most importantly during a five-month research trip in 2019, supported by a

Mitacs Fellowship. While there, I was generously hosted by the Geography Department at Trinity College Dublin, a most formative experience. My mentor there, Phil Lawton, was instrumental in helping me find my footing and orienting some of the tricky questions around the intersections of urban development and media/creative economies. He also got me a desk to work at amid some challenging Dublin real estate. It was at Trinity that I met my now frequent collaborator and good friend Patrick Bresnihan, whose influence (and citations) can be found throughout this book—I wouldn't be where I am without you. Thanks, comrade. I appreciate the colleagues and friends I made at Trinity, who I now have the privilege of seeing in Dublin: Cian O'Callaghan, Maedhbh Nic Lochlainn, Fiona Mc-Dermott, Paul O'Neill, and many others. I also thank all of my interlocutors across Ireland, too numerous to mention, who so generously donated their time to the research in this book.

This book started to become an actual book at McGill University in the Department of Art History and Communication Studies. Darin Barney has been an absurdly generous and supportive mentor and friend. His interest in my research and clarity of understanding around infrastructure and rurality has forever changed the direction of my career. I had the privilege of sharing an office with Jordan Kinder, where we were able to bounce ideas about media and the environment off each other over coffees and beers as we emerged from the pandemic into a brutal job market. The proposal for the book also took shape with insights from Jordan and Sophie Toupin. They and the rest of the Grierson Research Group were an incredibly warm presence for me during my time at McGill: Farah Atoui, Janna Frenzel, Burç Köstem, Laura Pannekoek, Rafico Ruiz, Malcolm Sanger, Hannah Tollefson, Ayesha Vemuri—thanks for being great. Thanks to the Fonds de Recherche du Québec - Société et Culture (FRQSC) for funding my time there.

The final stages of this book were prepared at University College Dublin in the School of Information and Communication Studies, where I've found an exceptionally welcoming and supportive environment. Thank you, Eugenia Siapera and Marguerite Barry for your fearless guidance as subsequent heads of school, allowing me time and flexibility for my various endeavors (and my challenging commute from Dundalk). Thanks to the friends, colleagues, and comrades at UCD who have listened to me, shared ideas, and/or given me hard advice as I finished the book amid unfolding crises: specifically, Pat Anthony, Jeremy Auerbach, Sharae Deckard, Treasa de Loughry, Liz Farries, Páraic Kerrigan, Lai Ma, Anne Mulhall, Dylan Murphy, Kalpana Shankar, and James Steinhoff. Other friends and comrades in Ireland have also indulged any of my complaints and, more importantly, helped work together toward a better and more nourishing academy here: V'cenza Cirefice, Louise Fitzgerald, Rory Rowan, Kate Stokes, Fiadh Tubridy, to name a few.

Kay Dickinson of Concordia University has always been a staunch comrade and supporter of this work from its early stages. Kay has influenced me toward scholarship and approaches far beyond what I had thought possible within our discipline. I can't explain how lucky I am to have her support and guidance, as a mentor and a friend. Charles Acland and Emer O'Toole were also so helpful and generous as I languished over this project in its early stages, offering insights and clarity when needed most. I'll also be forever thankful to the faculty in the Mel Hoppenheim School of Cinema for creating such a great environment for early career scholars: Masha Salazkina, Joshua Neves, Marc Steinberg, Haidee Wasson, Luca Caminati, and Martin Lefebvre especially. I benefited enormously from their support, as well as that of the Global Emergent Media Lab and the FRQSC. The frontline administrative staff in Mel Hoppenheim, especially Maggie Hallam and Ria Rombough, were enormously helpful and put out numerous fires for me during the COVID lockdowns in Montréal. I also want to acknowledge support from the School of Irish Studies, which provided funding, conversation, and a second disciplinary home for me at Concordia—Michael Kenneally, Matina Skalkogiannis, and Marion Mulvenna especially.

I've benefited enormously from the friendship and comradeship of an amazing array of media, geography, and environment scholars as this book has developed, who variously held me up, drunk beers with me, or just inspired me as the years have progressed. Fadi AbuNe'meh, Maria Corrigan, Giuseppe Fidotta, Derek Garcia, Sadie Gilker, Enrique Fibla Gutiérrez, Becky Holt, Philipp Keidl, David Leblanc, Sima Kokotović, Ylenia Olibet, Weixian Pan, Lola Rémy, Viviane Saglier, Joaquín Serpe, Egor Shmonin, Pat Smith, Sanaz Sohrabi, Nikola Stepić—thanks y'all. To my scattered environmental media and geography comrades—Patrick Bigger, Lisa Han, Alix Johnson, Dillon Mahmoudi, Rahul Mukherjee, Anne Pasek, Juliet Pinto, Nicole Starosielski, Hunter Vaughan, Julia Velkova—and anyone else I'm forgetting, thank you.

I want to thank my editor at Duke University Press, Courtney Berger, for her support of this project from the word go. Her reflective care and attention, along with early critical feedback orienting the volume's contribution, have made this the book that it is. Laura Jaramillo was incredibly generous and responsive in helping guide this through the many stages of review and last-minute changes and edits, as was the entire team at Duke. Finally, I must thank the two anonymous reviewers who provided sharp, kind feedback where the book needed it most.

And of course, I need to thank day one family and friends. Jesse, Erin, Mike, Alexa, Hugh, Dan M., Dan R., Blaise, Joey, Neha, Adrian, Eric, Adam, Paul, Aisling, Orla, Emily, Khai, and many others from Philly to Ireland have over the years helped me, housed me, drunk with me, and borne with me as I've taken on a career that makes it so I don't see any of you as much as I'd want to. My family has been my bedrock, especially during a pretty tough last few years. Mom, Liam, Katie, Marissa, Grandma Jean, and the Spontaks—thanks for everything. Shout out to Rex, my favorite transnational dog—our rescued chihuahua-miniature pinscher-dachshund mix living in Ireland via Alabama, Vermont, and Montréal. Finally, thank you to Eimear—you are my world, and you and your family (Rosie, Paul, Fionntan) have made moving to Ireland feel like I've come home. Tiocfaidh ár lá (and up the Reds).

Whoever I'm forgetting, you are hereby relieved of the burden of being acknowledged amid whatever mistakes I have made in this book, all of which are solely my own.

Arguments in chapter 2 of this book took shape in a very nascent state in Brodie (2019). An expanded analysis of the west Kerry screen tourism economy from chapter 3 appears in Brodie (2020b). Heavily revised portions of the following articles appear in chapter 4: Brodie (2020a, 2020c). All materials have been reproduced with permission.

This book is dedicated to my father, David Brodie, who passed away in 2022. To him I owe everything.

Introduction. A Rising Tide Lifts All Boats

1   Although "Ireland" constitutes all thirty-two counties on the island of Ireland, I am in this book using Ireland as shorthand for the Republic of Ireland, the twenty-six counties south of the border with the UK and Northern Ireland, which separates the historical "Free State" and the occupied six counties of the north.

2   The time period on which this book focuses is roughly 2007–19. However, this is not a strict timeline, and I make no pretense of doing "historical" work. Rather, methodologically and conceptually, I gather material based on moments, cases, and phenomena, gathered through fieldwork, policy research, discourse analysis, and other materials, that appear significant *within* this time frame in terms of media infrastructure and the built environment.

3   The "Celtic Phoenix" as an economic period has itself been characterized by critics as the more pejorative "leprechaun economics" (Allen 2019; Regan and Brazys 2018).

4   A terminological note: Throughout this book, I often use "FDI" as a broad term describing private inward investment at a large scale (e.g., multinationals), which is how the Industrial Development Authority (IDA) and other state and semi-state bodies use it (e.g., "FDI companies"). Such imprecise usage demonstrates the ubiquity and unquestioning commitment to the strategy, obviously painting over different power relations and

geographical distinctions between its various sources, while also suggesting still that they are primarily US-based. Where possible and relevant, I attend to the nuances of power and distinguish between this reductive definition of FDI and other forms of foreign investment, involvement, and intervention.

5   A glaring absence that many attuned readers will notice is the absence of an analysis of "legacy media" in Ireland, especially Ireland's robust public television and radio apparatuses as well as its complex landscape of print and related online news media. This is not by accident—ultimately, there are already many, much better and comprehensive, analyses of these systems, their infrastructures, and their effects than I could muster in these pages (see Barton and O'Brien 2004; McCarthy 2021; Mercille 2014; Morash 2010; Pettitt 2000; Phelan 2014). My analysis is specifically focused on the physical environments of media production and circulation as enablers of trans- and multinational capital—in short, the infrastructures supporting this media and the policy that most directly affects these environments. It probably, in this way, may also betray my background in film studies, as Ireland's "screen policy" today arises largely from its former Film Board and its emphasis on foreign productions. While the Broadcasting Authority of Ireland (BAI) and Raidió Teilifís Éireann (RTÉ) have played a significant role in film production and transnational partnerships of the "screen industries" in Ireland, that influence will be for future studies, within which I hope the frameworks of this book provide a useful tool in the arsenal of critique, especially in light of 2023's payment scandals at RTÉ.

6   The laboratory concept has been explored at length in both conservative "revisionist" histories as well as postcolonial theory—specifically, as a way to understand Ireland's unique role within Britain's colonial empire among its other holdings. Ireland's experimental character comes largely from the "political arithmetic" of William Petty's landscape improvement ideologies, which also in many ways formed the basis of the export of Irish engineers and administrators elsewhere in the colonies. (See Bhandar 2020; Bresnihan and Brodie 2025; Carroll 2007; Ohlmeyer 2023.)

7   Of course, as with most truncated histories such as this, there is much more texture to this story than I have the space to give here, as the focus of this book is contemporary. For a more complex infrastructural history of this period, with all of the dense navigations around postcoloniality and the challenges of delinking amid a changing world economic system, see Bresnihan and Brodie (2025).

8   I am referring here, of course, to Anderson's concept of nations as "imagined communities" (2006). The dominance of the "national question" in postcolonial Irish society has of course been discussed and com-

mented on at length by conservative and radical thinkers. In particular, Joe Cleary's idea of the revisionist construction of "de Valera's Ireland" as a place of exclusively regressive insularity is instructive—he makes the point that in spite of conservative social values instilled through governance, it was also a time of attempted, if imperfect and incomplete, national modernization. See Cleary (2002).

9   It should be noted that the above paragraph is a simplified version of a longer history of tension and experiment within the nation-building project, as a way to illustrate the mainstream cultural relevance of Lemass's economic policies more than any specifically textured historiography. This "origin story," with Lemass and his contemporary T. K. Whitaker as the heroes of Ireland's early economic liberalization, has an ongoing popular appeal among a liberal-conservative elite whose economic reality is dubious in the historiography. I reproduce it here for its cultural and metaphorical significance within the material enactment of the state's development programs across history and its pertinence to the centrality of FDI (see Bresnihan and Brodie 2025; McCabe 2020; O'Hearn 1990).

10   In Irish historical discourse, this is contested terrain, especially considering the ways in which "modernization" theory has been employed by revisionists to whitewash the crimes of the British Empire in Ireland and, by extension, to celebrate the innovative ethos of the Irish developmental state in developing its "industrialization by invitation" strategy in response to a chronically "underdeveloped" society. See Bresnihan and Brodie (2025).

11   More recently, the genocide in Gaza has in 2023–24 awakened both the most powerful progressive and anti-colonial solidarity organizing the island has seen since the northern civil rights movements in the 1960s–70s and the Dunnes workers anti-apartheid strikes in the 1980s, but it has also empowered reactionary, racist cross-border anti-immigrant and specifically anti-Muslim organizing with the English Defense League and Loyalists. These right-wing groups have capitalized on the ongoing austerity-era housing shortage to sow hatred among working-class and migrant communities across the island.

12   In the early 2020s alone, we saw both traditional examples like the "Irish" Twitter-maligned *Wild Mountain Thyme* (John Patrick Shanley, 2021) and *Irish Wish* (Janeen Damian, 2024), as well as a commercially successful unraveling of these dynamics in the Oscar-winning *The Banshees of Inisherin* (Martin McDonagh, 2022). The popular discourse that raged around each of these films points to the unresolved positioning of Ireland and Irish cultural identity within global media forms.

13   For this incredible logistical phrase, I owe gratitude to Kay Dickinson (see her recent book, 2024).

14   An adopted human geographer, at best.

15   Of the three, I have only lived in Dublin, where I lived for several months over the course of my research, and where I now work. Still—I would not propose that this gives me any particularly unique or privileged insight.

16   This suggestion of drawing what's "offscreen" into the kind of research we do in film and media studies is generative in its reference to the dynamics between the researcher and the media technologies used to do research. In media production, for a sound technician or editor, this would involve considering both "wild" sound—unedited, environmental sonic "atmosphere"—as well as the cleaner, edited, narrative sound of the day. These were sites and encounters that will remain, shall we say, "wild"—at the edges of the frame, "where the environment speaks back, where communication bows to intensity, where worlds collide, cultures clash, and things fall apart" (Halberstam and Nyong'o 2018, 454). Conceptions of politics are scrambled, frayed even, but are constituent and vibrantly of the place, the history, its relations, and its livelihoods.

17   Again, "post-," in spite of ongoing occupation in the six counties, and in spite of a more privileged position today within the imperial and extractive world order than other postcolonies in the Global South.

Chapter 1. Turbulent Waters

1   This US collapse may still be imminent, but empire remains unfortunately durable. As I will suggest throughout this book, what we saw in Ireland in the 2010s may be symptomatic of a death knell of a particular kind of "stable" neoliberal orthodoxy.

2   Ireland's classification is consistently under debate. During the Celtic Tiger, its destiny seemed one of a pure core country, allied with the dominant economic powers of the Global North. However, for much of its history, theorists have termed it "semi-peripheral," and since the financial crisis this seems to be a consensus. Its status as a tax haven also seems to suggest some enduring, or at least imaginative, peripherality—such "havens" are rarely considered a true feature of the core. See Nitzsche (2013); O'Boyle and Allen (2021); O'Hearn (2016).

3   As of the end of 2020, the limit point of this study, the "recovery" appeared to have peaked and was leading to another recession. However, recessionary austerity and the ongoing creep of privatization that has become the norm means that another crash will likely look different than the last.

4   The twentieth century in Ireland has been debated and written about in great detail, especially according to the ways in which the country's postcolonial condition—a European semi-periphery with a relatively successful economy that nonetheless emerged as a post-developmental

state from under the shadow of the British Empire—complicates many narratives, culturally, ethnically, economically, and politically, about postcoloniality and globalization. The examples are numerous, and ever-expanding, but largely come from the field of Irish studies, dominated by the intersections of history and literary studies. While I do not intend to rehearse these debates here, they form a crucial backdrop: Cleary (2002); Connolly (2004); Eagleton, Jameson, and Said (1997); Ignatiev (1995); Kitching (2015); Laird (2015); Lloyd (2003); Mulhall (2020); O'Hearn (2001); Ulin, Edwards, and O'Brien (2013).

5 The pharmaceutical sector in Ireland is enormously powerful, to a scale perhaps exceeding the financial and tech industries, and is central to much of the industrial development traced briefly here through the 1970s and 1980s. However, these companies and activities are beyond the scope of this book, by both disciplinary boundaries and necessity. Nonetheless, there have been many fascinating cases of urban and rural agitation against these multinationals (and their partners in the state), narrated in detail in Robert Allen's *No Global: The People of Ireland Versus the Multinationals* (2004), which have great resonance with case studies throughout this book.

6 This, as has been gestured toward, likely has as much to do with global spatial ordering and the racialization of Irish workers as "white" and the nation as a "white" European nation-state. O'Hearn notes how Irish manufacturing labor, in the early stages of liberalization, was not actually much cheaper than competitors (O'Hearn 2000), unlike the centrality of cheap, racialized migrant labor characterizing these strategies somewhere like East Asia (Ong 2006). Ireland's exceptional prosperity cannot be separated from these racialized dynamics (see Mulhall 2020).

7 In July 2020, this ruling was overturned as the EU ruled in favor of Apple and Ireland. As of the time of writing, Ireland is due to sign on to Organisation for Economic Co-operation and Development (OECD) tax harmonization at a 15 percent minimum, bringing an end to their 12.5 percent corporation tax rate that has characterized their contemporary economy. However, though 15 percent remains low, we will also see what further evasions may yet be innovated by the Irish state and its corporate partners.

8 This concept—"neoliberalism with Irish characteristics"—builds on David Harvey's theorization of "neoliberalism with Chinese characteristics" (2005), which articulates the way that an ostensibly socialist developmental state enacted certain limited neoliberal reforms. In addition, it was codeveloped alongside my work with Patrick Bresnihan in our book *From the Bog to the Cloud: Dependency and Eco-Modernity in Ireland* (2025), where we focus more specifically on the characteristics of Irish neoliberalism as it manifests in state and semi-state organizations to manage ecological contradictions.

9 In other work, I argue that this specific landscape of semi-states is central to the functional operation of "neoliberalism with Irish characteristics" (Bresnihan and Brodie 2025).

10 This is not to mention the actually existing digital divide in urban spaces as well, where the unhoused and other precarious or noncitizens are denied access by economic or racial barriers and discrimination.

11 Elizabeth Povinelli has written on "eventfulness" and "late liberal governance" in contexts of what she calls "economies of abandonment" in indigenous communities in Australia (2011). The eventful power of state and capital are contrasted with the noneventful and everyday suffering of abandoned populations. Rural Ireland is a quite different context, but the feelings of abandonment there remain significant.

12 As intervening years have made clear, this private sector inclusion—and its frequently populist orientation—has been a proving ground for often dangerous and haywire political associations, arguably a fertile field of recruitment for the far right in Ireland and elsewhere.

13 I am conscious of reproducing an academic tendency to expand the application of a term to its ultimate detriment—and infrastructure, as a term that has circulated through the various "material turns" over the last decade plus, is very much a saturated one. In this text, I have tried to ensure the following terminological specificity: 1) infrastructure as the physical material of economic production and circulation, and 2) the infrastructural character of industrial and social policies in producing the environments through which such material economies function. For more recent critiques and articulations of the infrastructural turn in media studies, see Barney (2021); Hesmondhalgh (2021); Parks, Velkova, and de Ridder (2023); Plantin and Punathambekar (2019).

14 The research for *Wild Tides* was performed almost exclusively before the "AI boom" in the early 2020s, so the specific applications have been historicized to account for the industrial shifts since then.

15 And in these ways, logistical thinking also demonstrates sites of developmental conflict and geopolitics. For example, a Chinese company was flagged for investing in a new Irish port project in Bremore (Tranum 2023).

16 In this, I join a chorus of media scholars focusing on these logistical media technologies. See Brodie, Han, and Pan (2019); Dickinson (2024); Hockenberry, Starosielski, and Zieger (2021); Rossiter (2016); Skvirsky (2020); among others.

17 And while this book focuses exclusively on the Republic of Ireland, we have to remember that a brutal civil war also raged north of the border in the occupied six counties from 1969 to 1998, which still had implications and effects for ordinary residents of the south—some of whom were participants in or victims of the conflict.

Chapter 2. Ghostly Currents and Creative Erosion

Arguments in this chapter took shape in a very nascent state in Brodie (2019).

1   Various interlocutors told me that counting the number of cranes over Dublin was akin to reading the stock market.

2   I spent hours wandering the Docklands and Poolbeg intermittently from 2017 to 2019, photographing and observing the constant and profound transformations occurring at often staggering speeds in areas of financialized property, in contrast to the industrial ruin and pollution in the "undeveloped" areas of the sector. Once, upon realizing my meandering through the buildings, water treatment sites, low-trafficked industrial roads, and paths had made me late for a meeting with my supervisor at Trinity, I realized I was an hour's walk from the nearest bus stop. I called a taxi to make the appointment. Chatting with the driver about why I was out there, he told me, by chance, that he frequently drove for film productions in the Dublin region. The Poolbeg Peninsula, he said, was no good. The infrastructure was lacking and out of date; it took too long to get out there from the city center; it was ugly. There were far better sites for film studios and production than Poolbeg, he argued, at least from a driver's perspective. The speculation on media as a new productive force in the area was just speculation—mixed-use development was the less risky venture, and would likely move ahead.

3   There have been a few horror films that represent the anxiety around suburban ghost estates especially. Lorcan Finnegan directed the short film *Foxes* (2011), a psychological thriller set in a ghost estate, and his feature *Vivarium* (2019), a *Twilight Zone*–inflected sci-fi set in a faceless suburban estate, debuted at Cannes. Finnegan has claimed that these films are influenced by the ghost estate phenomenon. In addition, I would argue that Ivan Kavanagh's 2014 film *The Canal* should be included in this cluster, a supernatural haunted house story that features key scenes in an abandoned, half-constructed house and whose plot elicits general fear and guilt manifested in the domestic built environment. Kristen Sheridan's *Dollhouse* (2012) deals with post-crisis class divisions of Dublin via housing, fictionally staging "inner-city" Dubliners breaking into, partying in, and wrecking an upscale suburban house.

4   A friend from Salthill, Galway, once told me a hilarious example from this period: Her older brother's high school graduation party was a multiple keg and bouncy castle affair, with a hundred attendees getting obliterated drunk in her parents' quiet suburban estate.

5   This is not to mention the erasure of social stratifications within the "we" pushed by public and state discourse, which mapped onto complexities of Ireland's racialized migrant and ethnic communities in the aftermath of the crisis (Titley 2015).

6 Reparative practices, both real and foreclosed in the recovery era, speak to a vibrant discourse around repair and care in media studies and STS (Graham and Thrift 2007; Jackson 2014; Mattern 2018).

7 By the time of my final visit in 2019, the Palás in Galway was fully operational. In general, there remains a huge problem with urban *and* rural vacancy in Ireland, which has been the subject of both fruitful housing organizing in the form of the island-wide CATU (Community Action Tenants Union) as well as growing anti-migrant sentiments around the use and mobilization of such vacancy. It is perhaps telling that the Palás cinema in Galway closed in late 2024 under market pressures, and the strongest advocate of its salvaging from dereliction is CATU's Galway branch (https://www.instagram.com/catugalway/p/DDw3RCQNjqS/?api=postMessage&img_index=1). See Gavin (2020); O'Callaghan and Di Feliciantonio (2023); O'Callaghan and Stokes (2021); Tubridy (2023).

8 Facebook, or Meta, has since moved to Ballsbridge, a more "old-money" neighborhood.

9 After nearly two decades of scholarly refutations and real-life examples of inflated property values and blue-collar displacement in cities, even Florida himself has admitted the limitations of this approach to actual urban development and analysis (2013).

10 While in the world of COVID-19 this has undoubtedly shifted, with the decline in office spaces and the imperatives of remote working, this has arguably crystallized into the city itself in the form of permanent austerity. As Dublin's cultural institutions have closed while expensive new apartment blocks, condos, and office buildings continue to proliferate, I'm sure it never had anything to do with "creativity" other than the destruction of truly alternative, let alone public and equitable, forms of social life and practice defined outside of a narrow vision of creative capital.

11 This definition notably excludes tourism, which I will address in chapter 3 as a significant omission from contemporary "creative industries" policy.

12 In the interest of disclosure, I was also an invited speaker on this walking tour/performance and want to give high credit and praise to my friend and colleague Paul O'Neill.

13 From a CFP for a conference on "Opaque Media" at UC Irvine in 2018: "media produced outside of commercial or academic channels and pertaining to the work of government agencies and departments, NGOs, private companies, consulting firms, and university centers (to name a few). Grey media encompasses the production and circulation—in print or electronic form—of a wide range of objects, including working papers, policy documents, technical specifications, environmental impact assessments, and contracts" (Opaque Media 2017).

14 In response to Mabo's displacement from the space, activist group Take Back the City occupied Airbnb's offices in 2018, arguing that companies like these were "colonizing" the city (Breaknews.ie 2018), which could describe both their business practices (which encourage landlords to buy up swaths of properties to rent to tourists for exorbitant prices) as well as their choice of headquarters.

15 After Notre Dame burned in Paris in spring 2019, the Irish History Podcast tweeted a popular sentiment, arguing that "If the Notre-Dame was in Dublin, there would be surveyors in the ruins right now figuring how they could convert it into a hotel or a hub for tax exempt tech startups [*sic*]" (Irish History Podcast 2019). This is both hilariously spot on and also not terribly far-fetched, as Boland's Quay, which may have once seemed an untouchable cultural landmark before its acquisition by NAMA in 2012 and Google in 2018, can show us.

16 Rose, Degen, and Melhuish are among the few to take seriously and try to incorporate geographical and industrial understandings of how this peculiar visual culture understands and organizes city space (in Doha, Qatar, 2014), as well as how such images are then felt, used, or experienced (Degen, Melhuish, and Rose 2017; Melhuish, Degen, and Rose 2016).

17 Having visited these offices in 2024, the flat, colorful "Google aesthetic" has subsumed the heritage of the space in unsurprising ways.

18 A friend in construction management once told me that most of his job is logistics, coordinating across providers, subcontractors, designers, engineers, architects, and the like. I've brought this insight to other construction workers, who have always agreed.

19 Until spring 2020, then-Taoiseach Leo Varadkar's Twitter bio read "Saviour of the Poolbeg Stacks. No kidding," reflecting their iconicity and also Varadkar's shameless pandering.

20 As of the conclusion of this study, there remained a hopeful thrust for development of the creative industries in these former industrial lands, with the largely vacant Pigeon House Power Station and Hotel also proposed for regeneration as a cultural and creative hub (Kelly 2019). All plans remained speculative and in contention, the industrial landmass intersecting with postindustrial, post-crisis economic visions of a new Dublin from the ruins of the old. And, of course, this is not to mention the implications of a climate-changed Dublin and the literal rising tides threatening to submerge Poolbeg and any "development opportunity" offered by it.

21 For example, they partially funded feminist artist Jesse Jones's *Tremble, Tremble* (2017), a monumental installation starring Irish performer Olwen Fouéré that premiered at the Venice Biennale. Relevant to the theoretical

strands of this chapter, the piece imagines a feminist legal and political order enriched by the unruly "multitude" in the form of a new body, In Utera Gigantae (E-Flux 2018), clearly influenced by autonomia political theory and action.

22　In the final 2019 planning document, all above-listed plans for *a* film studio—not necessarily Dublin Bay Studios—remained, as did a focus on the film, TV, and ICT sectors (Dublin City Council 2019). DPC, however, would not bend on their decision to leave the film studio out of the plans. As of 2019, Dublin Bay Studios still intended to develop elsewhere on the Poolbeg Peninsula, pending permission from DPC, trying to mimic Montréal and develop film production as well as a "VFX and video games cluster" (Slattery 2016), building off existing clusters in the Docklands and elsewhere in the city, expanding the remit of the film production infrastructure to include the wider "screen industries" (see chapter 3). Dublin Bay Studios has since relocated and redirected its energy to a planned media park in Grange Castle Business Park South—the site of a high concentration of data centers, aligning media and ICT sectors in a less central region of Dublin's sprawl (see chapter 4).

23　Lewis Doyle Singer, "The Two Chimneys," TikTok, December 14, 2024, 3 min. 11 sec., https://www.tiktok.com/@lewisdoylesinger/video/7448223380735610144.

24　This film was characterized as "haunting" by the *Irish Times* upon its release in 1996, emphasizing the almost surreally devastating nature of poverty and the history of women's exploitation, which continues to haunt Ireland (Donovan 1996).

Chapter 3. Waves of Austerity

An expanded analysis of the west Kerry screen tourism economy in this chapter appears in Brodie (2020b).

1　Sports media were frequently a source of secondary income for media workers I interacted with, especially the GAA in supporting its national and community remits.

2　As of the last time we spoke (February 2020), the specific grant we discussed was not forthcoming.

3　This applies to other cultural funding agencies as well, in particular the Arts Council, whose budget cuts have coincided with a familiar rhetorical shift toward profit-making since the crisis (Olsberg SPI with Nordicity 2017; Slaby 2011).

4　These components themselves, in true supply chain efficiency, may have eventually made it back to servers in the country's data processing industries (see chapter 4).

5   The Dublin Bay Studios venture has apparently now secured a site in the western outskirts of the city. Called the "Grange Castle Media Park," it is backed by a firm called Lens Media Ltd., consisting of Morris, Moloney, and American producer Gary Levinsohn. If it receives the proper permissions, it will share an IDA campus with Google's data center complex in Grange Castle Business Park South (Hamilton 2020).

6   This company name, a private equity firm under investor Joe Devine, is likely a reference to Sidney Olcott and the "O'Kalems," an early US film company that produced some of the first narrative moving images of Ireland in the early twentieth century. See Flynn (2011).

7   As if on cue to illustrate my point as I finish this typescript, Windmill Lane's postproduction studios closed their doors permanently in early 2025, citing "the erosion of international competitiveness of the Section 481 VFX film tax credit" (quoted in Tabbara 2025).

8   For a textbook example of economic impact justification for financialized media industrial policy, see PricewaterhouseCoopers (2020, 19–26).

9   SPVs, perhaps not coincidentally, reside in the finance sector, and were a contributing factor to the toxicity of the 2007–8 crash.

10   Ireland ranks high in abstract "soft power" metrics and is known for wielding this influence (see Brodie 2021; Charlemagne 2020).

11   This was also a yearly declining scheme, running from 2019 to 2023. It provided tax relief of an additional 5 percent in 2019 and 2020, which was extended to 2021 due to the COVID-19 pandemic. It then tapered off from 3 percent in 2022 to 2 percent in 2023, after which point it was nixed. There are proposals to extend the 5 percent scheme indefinitely, but with more stringently "regional" requirements. See Houses of the Oireachtas (2023).

12   At the risk of sounding old, I would also argue that Galway has lost a bit of its buzz because of many of the dynamics traced throughout this book, as well as post-COVID decline. That said, it is still a beautiful wee city.

13   It also points to the impossibility of maintaining these local cultures, as cultivated as they are by "starving artists," when the cost of living becomes prohibitive—arguably, it was easier to get by in Galway during the crisis than it would be now, "post-recovery."

14   This is, of course, related to long cultural histories of Irish emigration, and it is very normal for young people to emigrate to the United States, Canada, and Australia (and increasingly to places like the UAE), especially to work, either seasonally or on one- or two-year visas, as well as to other European countries.

15   Troy was shopped and became the location to shoot a program in Apple TV's initial content lineup (which became the Apple TV show *Founda-*

tion), much like the company's aborted data center in Athenry was reported to be building capacity for streaming traffic in Europe.

16   As an artist, she tried to secure arts funding to pursue activist projects, but had not been successful at this time in securing funding for any major political action—including a very funny détournement in which she tried to secure earmarked Dublin City Council arts funding to pursue a court case (Guinan 2020).

17   And resonant with our studies in chapter 2, the Creative Limerick—Connect to the Grid program encouraged "creative" and arts businesses to temporarily occupy vacant urban spaces, which are now for the most part returned to the "normal" property market (Limerick.ie n.d.).

18   For a full analysis of this screen tourism economy in west Kerry, see Brodie (2020b).

19   In a conversation with Irish film scholar Pat Brereton, he suggested simply standing at the boats in Portmagee and surveying each patron to ask how big a factor *Star Wars* was in their visit to Skellig Michael. Without Fáilte Ireland funding, however, I was hesitant to perform such an instrumental chore.

20   The scheme was time-capped, going down a percentage point each year, and had been completely phased out by the time of this book's completion.

21   Anecdotally, friends and I tried to get a place during the Dingle Arts Festival and would have had to stay up to a thirty-minute drive away from the town in a smaller neighboring community across or off the peninsula. It is widely known that if you don't book accommodation long before a bank holiday weekend, Dingle is an impossible bet.

22   This political economy of global production extends beyond Hollywood, of course. Bollywood blockbuster *Ek Tha Tiger* (Kabir Khan, 2012), for example, capitalized on a growing Indian diasporic connection in Ireland by featuring a plot largely centered on a researcher at Trinity College, Dublin, boasting an extended song sequence complete with hurls, Kilkenny GAA jerseys, a parade of green leprechaun hats, and bagpipes. It also received generous S481 benefits at the time. See Rampazzo (2017).

23   I have elsewhere referred to this as a "branding from below," building on Verónica Gago's concept of "neoliberalism from below" (2017). See Brodie (2020b).

24   In a potent example, in 2010 while filming the Italian/French/Irish co-production *This Must Be the Place* (Paolo Sorrentino, 2011) in Ireland, Sorrentino as an EU citizen working in Ireland would justifiably have passed the "cultural test" as "eligible spend" for the S481, whereas US actor Sean Penn would not have. However, under stipulations begun

in 2015, under what is called the "Tom Cruise clause" (Murphy and O'Brien 2015), as talent working *in* Ireland, Penn's wages would have been eligible for a hefty tax break by the company's producers (up to €50 million). In effect, as Murphy and O'Brien argue, a production could under these rules technically employ *no* Irish-based workers or unions whatsoever and "and with few obligations to workers beyond the minimum wage and health, safety and working time standards" (2015). For more analysis of quantification and cinematic identity, see also Tracy and Flynn (2016).

25 For an entertaining background, see *It Came from Connemara* (Brian Reddin, 2014).

26 While this study focuses on precarity as a shared condition of work across lines of race, gender, sexuality, and dis/ability, exploitation is nonetheless concentrated more severely and more violently on racialized subjects, women, members of the LGBTQ+ community, people with dis/ability, and other marginalized groups in Irish society. The work of Aphra Kerr (with Preston, 2001), Anne O'Brien (2014), Kerrigan (with O'Brien, 2020), Zélie Asava (2013), as well as Screen Ireland's recent reports and programs designed to achieve gender equity and overall diversity (see Fís Éireann/Screen Ireland 2020), provide the groundwork for future studies on these key areas, which are essential for imagining and creating more just cultural and economic futures for Irish media industries.

Chapter 4. Storm Clouds

Portions of this chapter first appeared in different forms in Brodie (2020a, 2020c).

1 In the interest of privacy, I have adopted a pseudonym for the Athenry for Apple member in this chapter.

2 There are, of course, many problems surrounding industrial-scale monoculture tree plantations, including significant effects on local communities, soil quality, agriculture, biodiversity, and wildlife (Fitzgerald 2023).

3 After a later email exchange with Daly, I stand by this assessment.

4 UK-based artist Matt Parker (2019, n.d.) created an installation about the thickness of these kinds of feelings and cultural elements, echoing the immanent critique of someone like Elizabeth Povinelli and the "thickness" of social projects and subjectivities in late liberalism (2011, 6).

5 For full analysis of the Athenry saga and its relations to populism and political subjectivity, see Brodie (2020c).

6 These real and projected numbers have continued to rise. In 2022, data centers used 18 percent of Ireland's electricity on a daily basis, a number

expected to rise to 32 percent by 2026 (see Brodie 2024b; Jain 2024). In 2023, the number had already risen to 21 percent.

7   Energy shortages in the Dublin region, predicted by projections (see Lima 2019), came to pass in 2021 and 2022, as a lack of significant state investment in new, nonallocated infrastructure in the coming years has coincided with data centers' disproportionate demands on the grid. There has been a good deal of public waffling over whether or not limitations imposed by EirGrid in 2021 on grid connections constitute a "moratorium" on data centers in the Dublin region (Butler 2023). Data centers seem to be being built relatively unabated (see Brodie 2023, 2024b).

8   The canal itself used to map an industrial corridor between Dublin and the River Shannon in the west of Ireland. Now, more ironically, the no-longer-functional canal connects this particular business park with Grand Canal Dock, the heart of the Dublin Docklands, where tech companies such as Facebook, Google, and Airbnb have their expansive European headquarters, as detailed in chapter 2. The commuter train lines heading out to Dublin's western suburbs and beyond to Galway (through Athenry) and Limerick still roughly follow along this original transport infrastructure, crisscrossing it through the small towns and out through the Midlands, and demonstrating the historical longevity of infrastructural path-dependence and the continuous activation of existing spaces and routes for circulation.

9   In 2023, as I complete this book, that number is still rising, in spite of new limitations on growth imposed in 2021–22. According to energy solutions company Bitpower, there were eighty-two operational data centers in Ireland in 2023.

10   While not a moratorium, a 2021 report from the Commission for Regulation of Utilities (CRU) indicated that rolling blackouts would begin in Dublin without a rethinking of data-center development and encouraging construction in more rural areas (CRU 2021), which led EirGrid in late 2021 to impose curbs on data-center grid connections in Dublin for five years (Swinhoe 2022).

11   Laura Marks, in a short piece, established that streaming video accounted for 60 percent of the world's data traffic and 1 percent of total greenhouse gas emissions at this time (2020).

12   This is not an attempt to erase the existing labor at these sites, including the continuing maintenance and security labor of blue-collar workers. Julia Velkova has done exceptional work on these sorts of labor (2020; Mayer and Velkova 2023), as have Jenna Burrell (2020) and Vicki Mayer (2023). However, it remains true that the number of workers represented within these enormous infrastructures is proportionally quite low in comparison to the scale of their operations, capital investment, and infrastructural affordances.

13   Coillte's project page still existed as of 2020, two years after Apple's withdrawal (Coillte n.d.).

14   This book was drafted before the recent "AI boom," so AI has not been included as a primary structural factor in the Irish industry. Up to 2020, it did not appear to be robustly on the radar of the data center industry in the country, at least in any terms that were shaping their plans.

15   Public agitation toward imperial routes of natural resource extraction has risen in Ireland. US-based academic Aviva Chomsky has criticized ESB for using coal from open-pit mines in Colombia (O'Sullivan 2019).

16   Lisa Parks has written evocatively on the environmental and visibility politics of "antenna trees," the fake trees made of telecommunications towers in forested areas (2009).

17   It should be noted that of course "climate" is also something that is universalized and globalized, and in fact climate science is frequently articulated as a medium and historical fulcrum of global perception. See Edwards (2010).

18   The 1970s leftist group the Weather Underground usefully used a similar meteorological association, via Bob Dylan—"You don't need a weatherman to know which way the wind blows"—suggesting that certain actions and reactions simply cannot be encapsulated and captured by corporate and state forms of vision and control, and must be made fluctuatingly actionable and reacted to. Hito Steyerl, in the video *Liquidity Inc.* (2014), similarly associates the liquidity of contemporary circulatory capitalism and unpredictable movements, giving ironic weather forecasts as "The Weather Underground."

19   As noted above, Apple and Ireland were firm partners on this legal battle for corporation tax evasion. However, Ireland's sign-on to OECD tax harmonization (at 15 percent) may have slightly lessened its competitive advantage on this front, although its effective tax rate remains comparatively very low.

20   Hogan, for example, has researched how DNA storage may be the new frontier of data storage for tech, demonstrating further entanglements with nature (2018b). Quantum computing is another speculative area of research in this area.

Conclusion. Calmer Seas

1   We can think, for example, of the Land Wars of the late 1870s and early 1880s, or of the enduring legacy of the "land question" in postcolonial and environmental organizing. For more expansion on this, see Bresnihan and Brodie (2025).

REFERENCES

Acland, Charles R. 2018. "An Empire of Pixels: Canadian Cultural Enterprise in the Visual Effects Industry." In *Reading Between the Borderlines: Cultural Production and Consumption Across the Canada-US Border*, edited by Gillian Roberts, 143–70. Montreal: McGill-Queen's University Press.

Alimahomed-Wilson, Jake, and Immanuel Ness. 2018. "Introduction: Forging Workers' Resistance Across the Global Supply Chain." In *Chokepoints: Logistics Workers Disrupting the Global Supply Chain*, 1–15. London: Pluto Press.

Allen, Mary. 2019. "Celtic Phoenix Rising on Data Centre Wings." *InsightaaS*, October 12. https://insightaas.com/celtic-phoenix-rising-on-data-centre-wings/.

Allen, Robert. 2004. *No Global: The People of Ireland Versus the Multinationals*. London: Pluto Press.

Amoore, Louise. 2018. "Cloud Geographies: Computing, Data, Sovereignty." *Progress in Human Geography* 42, no. 1: 4–24.

An Bord Pleanála. 2021. "S. 4(1) of Planning and Development (Housing) and Residential Tenancies Act 2016: Inspector's Report ABP-308917-20." April 6. https://www.pleanala.ie/anbordpleanala/media/abp/cases/reports/308/r308917.pdf.

Anand, Nikhil. 2019. "Leaking Lines." In *Infrastructure, Environment, and Life in the Anthropocene*, edited by Kregg Hetherington, 149–68. Durham, NC: Duke University Press.

Anand, Nikhil, Akhil Gupta, and Hannah Appel, eds. 2018. *The Promise of Infrastructure*. Durham, NC: Duke University Press.

Anderson, Benedict. 2006 (1983). *Imagined Communities: Reflections on the Origin and Spread of Nationalism*, rev. ed. Brooklyn: Verso.

Anex, Valérie. 2011. "Ghost Estates." Accessed July 29, 2020. http://www.valerieanex.com/index.php/ghost-estates/.

Apprich, Clemens, and Ned Rossiter. 2016. "Sovereign Media, Critical Infrastructures, and Political Subjectivity." In *Across and Beyond: A Transmediale Reader on Post-Digital Practices, Concepts, and Institutions*, edited by Ryan Bishop, Kristoffer Gansing, Jussi Parikka, and Elvia Wilk, 270–83. Berlin: Sternberg Press.

Arboleda, Martín. 2020. *Planetary Mine: Territories of Extraction Under Late Capitalism*. Brooklyn: Verso.

Arrighi, Giovanni. 2010 (1994). *The Long Twentieth Century: Money, Power, and the Origins of Our Times*. Brooklyn: Verso.

Asava, Zélie. 2013. *The Black Irish Onscreen: Representing Black and Mixed-Race Identities on Irish Film and Television*. New York: Peter Lang.

Bailey, Sean. 2019. "Conversation with Paddy Breathnach." Public talk, Lighthouse Cinema, Dublin, Ireland, March 20.

Ballestero, Andrea. 2019. "The Underground as Infrastructure? Water, Figure/Ground Reversals, and Dissolutions in Sardinal." In *Infrastructure, Environment, and Life in the Anthropocene*, edited by Kregg Hetherington, 17–44. Durham, NC: Duke University Press.

Balsom, Erika. 2009. "A Cinema in the Gallery, a Cinema in Ruins." *Screen* 50, no. 4: 411–27.

Barney, Darin. 2011. "To Hear the Whistle Blow: Technology and Politics on the Battle River Branch Line." *Topia: Canadian Journal of Cultural Studies* 25: 5–28.

Barney, Darin. 2021. "Infrastructure and the Form of Politics." *Canadian Journal of Communication* 46, no. 2: 225–46.

Barton, Ruth. 2004. *Irish National Cinema*. New York: Routledge.

Barton, Ruth. 2015. "Between Modernity and Marginality: Celtic Tiger Cinema." In *From Prosperity to Austerity*, edited by Eamon Maher and Eugene O'Brien, 218–29. Manchester University Press.

Barton, Ruth. 2018. "Behind Closed Doors: Middle Class Suburbia and Contemporary Irish Cinema." In *Imagining Suburbia in Literature and Culture*, edited by Eoghan Smith and Simon Workman, 191–208. London: Palgrave Macmillan.

Barton. Ruth. 2019. "The Force Meets Kittiwake: Shooting *Star Wars* on Skellig Michael." In *The Routledge Handbook of Popular Culture and Tourism*, edited by Christine Lundberg and Vassilios Ziakas, 300–310. New York: Routledge.

Barton, Ruth, and Denis Murphy. 2020. *Ecologies of Cultural Production: Career Construction in Irish Film, TV Drama and Theatre*. Dublin: Creative Ireland. https://www.creativeireland.gov.ie/app/uploads/2020/03/ECP.pdf.

Barton, Ruth, and Harvey O'Brien, eds. 2004. *Keeping It Real: Irish Film and Television*. London: Wallflower Press.

Baxtel. n.d. "Dublin Data Center Market." Accessed July 29, 2020. https://baxtel .com/data-center/dublin.

Benjamin, Walter. 2008. "The Work of Art in the Age of Its Technological Re-producibility (Second Version)." In *The Work of Art in the Age of Its Techno-logical Reproducibility and Other Writings on Media*, translated by Edmund Jephcott, Rodney Livingstone, Howard Eiland, et al., edited by Michael W. Jennings, Brigid Doherty, and Thomas Y. Levin, 19–55. Cambridge, MA: Harvard University Press.

Berardi, Franco. 2009. *The Soul at Work: From Alienation to Autonomy*. Los An-geles: Semiotext(e).

Berlant, Lauren. 2011. *Cruel Optimism*. Durham, NC: Duke University Press.

Berlant, Lauren. 2016. "The Commons: Infrastructures for Troubling Times." *Environment and Planning D: Society and Space* 34, no. 3: 393–419.

Bernes, Jasper. 2013. "Logistics, Counterlogistics and the Communist Pros-pect." *Endnotes* 3. http://endnotes.org.uk/en/jasper-bernes-logistics -counterlogistics-and-the-communist-prospect.

Bhandar, Brenna. 2020. "Lost Property: The Continuing Violence of Improve-ment." *Architectural Review*, October 8. https://www.architectural-review .com/essays/lost-property-the-continuing-violence-of-improvement.

Bodkin, Peter. 2017. "Data Centres Are Making Energy Demands Shoot Up in Ireland." *TheJournal.ie*, May 6. http://www.thejournal.ie/data-centres -ireland-3-3374827-May2017/.

Bodkin, Peter. 2018. "Ireland's Biggest Film Studio Has Been Sold to a Private Equity Investor." *Fora.ie*, March 23. https://fora.ie/ardmore-studios-sold -3920291-Mar2018/.

Boland's Quay. n.d. *Reaching New Heights in a Prime Location*. Accessed July 29, 2020. http://bolandsquay.com/brochure.php.

Bousquet, Marc. 2008. "Students Are Already Workers." In *How the University Works: Higher Education and the Low Wage Nation*, 126–56. New York: NYU Press.

Bowker, Geoffrey C., and Susan Leigh Star. 1999. *Sorting Things Out: Classifica-tion and Its Consequences*. Cambridge, MA: MIT Press.

Boyd, Gary A., and John McLaughlin. 2015. "Introduction." In *Infrastructure and the Architectures of Modernity in Ireland 1916–2016*, edited by Gary A. Boyd and John McLaughlin, 1–8. Burlington, VT: Ashgate.

Breaknews.ie. 2018. "Take Back the City Occupation of AirBnb Offices Ends as Activists Blame Firm for Exacerbating the Housing Crisis." October 13. https://www.breakingnews.ie/ireland/take-back-the-city-occupation-of -airbnb-offices-ends-as-activists-blame-firm-for-exacerbating-the-housing -crisis-875583.html.

Bresnihan, Patrick. 2016. "The Bio-Financialization of Irish Water: New Advances in the Neoliberalization of Vital Services." *Utilities Policy* 40: 115–24.

Bresnihan, Patrick, and Patrick Brodie. 2021. "New Extractive Frontiers in Ireland and the Moebius Strip of Wind/Data." *Environment and Planning E: Nature and Space* 4, no. 4: 1645–64.

Bresnihan, Patrick, and Patrick Brodie. 2023. "Data Sinks, Carbon Services: Waste, Storage, and Energy Cultures on Ireland's Peat Bogs." *New Media and Society* 25, no. 2: 361–83.

Bresnihan, Patrick, and Patrick Brodie. 2025. *From the Bog to the Cloud: Dependency and Eco-Modernity in Ireland.* Bristol: Bristol University Press.

Bresnihan, Patrick, and Arielle Hesse. 2021. "Political Ecologies of Infrastructural and Intestinal Decay." *Environment and Planning E: Nature and Space* 4, no. 3: 778–98.

Brodie, Patrick. 2019. "Seeing Ghosts: Crisis, Ruin, and the Creative Industries." *Continuum* 3, no. 5: 525–39.

Brodie, Patrick. 2020a. "Climate Extraction and Supply Chains of Data." *Media, Culture and Society* 42, nos. 7–8: 1095–1114.

Brodie, Patrick. 2020b. "*Star Wars* and the Production and Circulation of Culture Along Ireland's Wild Atlantic Way." *Journal of Popular Culture* 53, no. 3: 667–95.

Brodie, Patrick. 2020c. "'Stuck in Mud in the Fields of Athenry': Apple, Territory, and Popular Politics." *Culture Machine* 19.

Brodie, Patrick. 2021. "Hosting Cultures: Placing the Global Data Center 'Industry.'" *Canadian Journal of Communication* 46, no. 2: 151–76.

Brodie, Patrick. 2023. "Data Infrastructure Studies on an Unequal Planet." *Big Data and Society* 10, no. 1: 1–14. https://doi.org/10.1177/20539517231182402.

Brodie, Patrick. 2024a. "Foreign Direct Investment and the Facilitation of Circulation in Irish Film Policy." *Irish Journal of Arts Management and Cultural Policy* 10, no. 2: 82–99.

Brodie, Patrick. 2024b. "Smarter, Greener Extractivism: Digital Infrastructures and the Harnessing of New Resources." *Information, Communication and Society* 28, no. 6: 1061–80. https://doi.org/10.1080/1369118X.2024.2341013.

Brodie, Patrick, and Darin Barney, eds. 2026. *Media Rurality.* Durham, NC: Duke University Press.

Brodie, Patrick, Lisa Han, and Weixian Pan. 2019. "Becoming Environmental: Media, Logistics, and Ecological Change." *Synoptique: An Online Journal of Film and Moving Image Studies* 8, no. 1: 6–13.

Brodie, Patrick, and Julia Velkova. 2021. "Cloud Ruins: Ericsson's Vaudreuil-Dorion Data Center and Infrastructural Abandonment." *Information, Communication and Society* 24, no. 6: 869–85.

Burke, Ceimin. 2019. "Government Signs Off on Contract for Controversial €3 Billion National Broadband Plan." *Thejournal.ie*, November 19. https://www.thejournal.ie/government-sign-national-broadband-plan-4896090-Nov2019/.

Burrell, Jenna. 2019. "Are Rural Data Center Jobs 'Good' Jobs?" *Global Media Technologies and Cultures Lab Blog*, September 3. http://globalmedia.mit .edu/2019/09/03/are-rural-data-center-jobs-good-jobs/.

Burrell, Jenna. 2020. "On Half-Built Assemblages: Waiting for a Data Center in Prineville, Oregon." *Engaging Science, Technology, and Society* 6: 283–305.

Burrington, Ingrid. 2016. *Networks of New York: An Illustrated Field Guide to Urban Internet Infrastructure*. Brooklyn: Melville House.

Butler, Georgia. 2023. "Ireland Isn't Going to Limit Data Centers Despite High Energy Use." *Data Centre Dynamics*, June 13. https://www.datacenterdynamics .com/en/news/ireland-isnt-going-to-limit-data-centers-despite-high -energy-use/.

Byrne, Michael. 2016. "'Asset Price Urbanism' and Financialization After the Crisis: Ireland's National Asset Management Agency." *International Journal of Urban and Regional Research* 40, no. 1: 31–45.

Caldwell, John Thornton. 2008. *Production Culture: Industrial Reflexivity and Critical Practice in Film and Television*. Durham, NC: Duke University Press.

Caldwell, John Thornton. 2013. "Stress Aesthetics and Deprivation Payroll Systems." In *Behind the Screen: Inside European Production Cultures*, edited by Petr Szczepanik and Patrick Vonderau, 91–111. New York: Palgrave Macmillan.

Caldwell, John Thornton. 2023. *Specworld: Folds, Faults, and Fractures in Embedded Creator Industries*. Oakland: University of California Press.

Callanan, Brian. 2000. *Ireland's Shannon Story: Leaders, Visions, and Networks: A Case Study of Local and Regional Development*. Dublin: Irish Academic Press.

Carolan, Mary. 2017a. "Apple Given Go-Ahead to Start €850m Data Centre in Athenry." *Irish Times*, October 12. https://www.irishtimes.com/business /technology/apple-given-go-ahead-to-start- 850m-data-centre-in- athenry-1.3253502.

Carolan, Mary. 2017b. "Storm Delays Apple Data Centre Appeal Hearing." *Irish Times*, October 16. https://www.irishtimes.com/business/technology /storm-delays-apple-data-centre-appeal-hearing- 1.3257635.

Carroll, Patrick. 2007. *Science, Culture, and Modern State Formation*. Berkeley: University of California Press.

Carroll, Rory. 2020. "Why Ireland's Data Centre Boom Is Complicating Climate Efforts." *Irish Times*, January 6. https://www.irishtimes.com/business /technology/why-ireland-s-data-centre-boom-is-complicating-climate -efforts-1.4131768.

Carruth, Allison. 2014. "The Digital Cloud and the Micropolitics of Energy." *Public Culture* 26, no. 2: 339–64.

Carse, Ashley, and David Kneas. 2019. "Unbuilt and Unfinished: The Temporalities of Infrastructure." *Environment and Society: Advances in Research* 10: 9–28.

Carswell, Simon. 2017. "Weekend Read: The Ghost Estates That Still Haunt Ireland." *Irish Times*, August 12. https://www.irishtimes.com/life-and-style/people/weekend-read-the-ghost-estates-that-still-haunt-ireland-1.3181498.

Carville, Justin. 2013. "Terra Infirma: The Territory of the Visible and the Writing of Ireland's Visual Culture." In *Viewpoints: Theoretical Perspectives on Irish Visual Texts*, edited by Claire Bracken and Emma Radley, 13–28. Cork: Cork University Press.

Central Statistics Office. 2022. "Data Centres Metered Electricity Consumption 2020." An Phríomh-Oifig Staidrimh/Central Statistics Office, January 20. https://www.cso.ie/en/releasesandpublications/ep/p-dcmec/datacentresmeteredelectricityconsumption2020/keyfindings/.

Chakrabarty, Dipesh. 2000. *Provincializing Europe: Postcolonial Thought and Historical Difference*. Princeton, NJ: Princeton University Press.

Chakrabarty, Dipesh. 2009. "The Climate of History: Four Theses." *Critical Inquiry* 35, no. 2: 197–222.

Charlemagne. 2020. "How Ireland Gets Its Way: An Unlikely Diplomatic Superpower." *The Economist*, July 18. https://www.economist.com/europe/2020/07/18/how-ireland-gets-its-way.

Chatterjee, Partha. 2004. *The Politics of the Governed: Reflections of Popular Politics in Most of the World*. New York: Columbia University Press.

Childs, Quincy. 2022. "'This Has Nothing to Do with Clouds': A Decolonial Approach to Data Centers in the Node Pole." *Commonplace*, June 29. https://doi.org/10.21428/6ffd8432.59c985d5.

Chua, Charmaine, Martin Danyluk, Deborah Cowen, and Laleh Khalili. 2018. "Introduction: Turbulent Circulation: Building a Critical Engagement with Logistics." *Environment and Planning D: Society and Space* 36, no. 4: 617–29.

Chung, Hye Jean. 2012. "Media Heterotopia and Transnational Filmmaking: Mapping Real and Virtual Worlds." *Cinema Journal* 51, no. 4: 87–109.

Cleary, Joe. 2002. "Misplaced Ideas? Locating and Dislocating Ireland in Colonial and Postcolonial Studies." In *Marxism, Modernity and Postcolonial Studies*, edited by Crystal Bartolovich and Neil Lazarus, 101–24. Cambridge: Cambridge University Press.

Climate Home News. 2017. "'Tsunami of Data' Could Consume One Fifth of Global Electricity by 2025." *Guardian*, December 11. https://www.theguardian.com/environment/2017/dec/11/tsunami-of-data-could-consume-fifth-global-electricity-by-2025.

Clover, Joshua. 2016. *Riot. Strike. Riot: The New Era of Uprisings*. Brooklyn: Verso.

Coillte. n.d. "Apple Derrydonnell." Accessed July 29, 2020. https://www.coillte.ie/our-business/our-projects/apple-derrydonnell/.

Collins, Patrick, and Seamus Grimes. 2011. "Cost-Competitive Places: Shifting Fortunes and the Closure of Dell's Manufacturing Facility in Ireland." *European Urban and Regional Studies* 18, no. 4: 406–26.

Connolly, Maeve. 2004. "Sighting an Irish Avant-Garde in the Intersection of Local and International Film Cultures." *Boundary 2* 31, no. 1: 243–65.

Connolly, William. 2017. *Facing the Planetary: Entangled Humanism and the Politics of Swarming*. Durham, NC: Duke University Press.

Cooper, Melinda. 2010. "Turbulent Worlds: Financial Markets and Environmental Crisis." *Theory, Culture and Society* 27, no. 2–3: 167–90.

Couldry, Nick, and Ulises A. Mejias. 2019. *The Costs of Connection: How Data Is Colonizing Human Life and Appropriating It for Capitalism*. Stanford, CA: Stanford University Press.

Coulter, Colin. 2015. "Ireland Under Austerity: An Introduction to the Book." In *Ireland Under Austerity: Neoliberal Crisis, Neoliberal Solutions*, edited by Colin Coulter and Angela Nagle, 1–43. Manchester: Manchester University Press.

Cowen, Deborah. 2014. *The Deadly Life of Logistics: Mapping Violence in Global Trade*. Minneapolis: University of Minnesota Press.

Cram, E. 2021. *Violent Inheritance: Sexuality, Land, and Energy in Making the North American West*. Oakland: University of California Press.

Crosson, Seán. 2003. "Vanishing Point: An Examination of Some Consequences of Globalisation for Contemporary Irish Film." *E-Keltoi: Journal of Interdisciplinary Celtic Studies* 2: 1–23.

Crowe, Horwath. 2017. *Final Report in Respect of a Strategy for the Development of Skills for the Audiovisual Industry in Ireland*. https://www.screenireland.ie/images/uploads/general/AV_Skills_Strategy_Report.pdf.

CRU. 2021. "CRU Direction to the System Operators Related to Data Centre Grid Connection Processing." Commission for Regulation of Utilities, November 23. https://cruie-live-96ca64acab2247eca8a850a7e54b-5b34f62.divio-media.com/documents/CRU21124-CRU-Direction-to-the-System-Operators-related-to-Data-Centre-grid-connection-.pdf.

Curtin, Michael. 2016. "Regulating the Global Infrastructure of Film Labor Exploitation." *International Journal of Cultural Policy* 22, no. 5: 673–85.

Curtin, Michael, Jennifer Holt, and Kevin Sanson, eds. 2014. *Distribution Revolution: Conversations About the Digital Future of Film and Television*. Berkeley: University of California Press.

Curtin, Michael, and Kevin Sanson, eds. 2016. *Precarious Creativity: Global Media and Local Labor*. Berkeley: University of California Press.

Curtin, Michael, and Kevin Sanson, eds. 2017. *Voices of Labor: Creativity, Craft, and Conflict in Global Hollywood*. Berkeley: University of California Press.

Dalton, David, and Warren Marmion. 2016. "Waves of Disruption: The Future of Ireland's Financial Services Sector." Dublin: Deloitte & Touche. https://www.readkong.com/page/waves-of-disruption-the-future-of-ireland-s-financial-8742566.

Daly, Gavin. 2019. "Apple Seeking a Buyer for Athenry Data Centre Site." *The Times*, October 13. https://www.thetimes.co.uk/article/apple-seeking-a

-buyer-for-athenry-data-centre-site-d7vs0md8m?fbclid=IwAR2iS8593jCx-LpKhNeRvsOJ5ikVQe4Yh8AIpv3OqNB9fSwjBRL6OciKaD6o.

Danyluk, Martin. 2018. "Capital's Logistical Fix: Accumulation, Globalization, and the Survival of Capitalism." *Environment and Planning D: Society and Space* 36, no. 4: 630–47.

D'Avella, Nicholas. 2019. *Concrete Dreams: Practice, Value, and Built Environments in Post-Crisis Buenos Aires*. Durham, NC: Duke University Press.

Davis, Mike, and Daniel Bertrand Monk. 2007. "Introduction." In *Evil Paradises: Dreamworlds of Neoliberalism*, edited by Mike Davis and Daniel Bertrand Monk. New York: New Press.

Deckard, Sharae. 2016. "World-Ecology and Ireland: The Neoliberal Ecological Regime." *Journal of World-Systems Research* 22, no. 1: 145–76.

Degen, Monica, Clare Melhuish, and Gillian Rose. 2017. "Producing Place Atmospheres Digitally: Architecture, Digital Visualisation Practices and the Experience Economy." *Journal of Consumer Culture* 17, no. 1: 3–24.

Department for Arts, Culture, and the Gaeltacht. 2011. *Creative Capital: Building Ireland's Audiovisual Creative Economy*. https://www.screenireland.ie/images/uploads/general/Creative_Capital_Web.pdf.

Department of Culture, Heritage and the Gaeltacht. 2015. "Projects and Programmes." July 7. https://www.chg.gov.ie/arts/creative-arts/projects-and-programmes/.

Department of Culture, Heritage and the Gaeltacht. 2017. *Audiovisual Action Plan: Creative Ireland Pillar 4*. https://www.chg.gov.ie/app/uploads/2018/06/audiovisual-action-plan.pdf.

DeWaard, Andrew. 2020. "Financialized Hollywood: Institutional Investment, Venture Capital, and Private Equity in the Film and Television Industry." *JCMS: Journal of Cinema and Media Studies* 59, no. 4.

Dickinson, Kay. 2024. *Supply Chain Cinema: Producing Global Film Workers*. London: BFI Publishing.

Dingle Slea Head Tours. n.d. "Dingle Peninsula Star Wars Tour." Accessed July 29, 2020. https://www.dinglesleaheadtours.com/dingle-peninsula-star-wars-tour/.

Donovan, Katie. 1996. "Haunted by the Poorhouse." *Irish Times*, March 30. https://www.irishtimes.com/news/haunted-by-the-poorhouse-1.37293.

Dowling, Tom. 2011. "Could We Transform a Ghost Estate into a Global Film Streetscape?" *Tom Dowling*, May 31. https://tomtdowling.wordpress.com/2011/05/31/could-we-transform-a-ghost-estate-into-a-global-film-streetscape/.

Dublin City Council. 2017a. "Chief Executive's Report on Submissions Received on the Proposed Material Alterations to the Draft Poolbeg West Planning Scheme." September 20. https://councilmeetings.dublincity.ie/mgConvert2PDF.aspx?ID=12724.

Dublin City Council. 2017b. *Poolbeg West Planning Scheme: Interim Publication.* https://www.dublincity.ie/sites/default/files/content/Planning/OtherDevelopmentPlans/Documents/Poolbeg%20West%20Planning%20Scheme%20(Interim%20%20Document).PDF.

Dublin City Council. 2019. *Poolbeg West Planning Scheme.* April. https://www.dublincity.ie/sites/default/files/content/Planning/OtherDevelopmentPlans/SpecialPlanningControlSchemes/Documents/Poolbeg_West_SDZ_Planning_Document.pdf.

Eagleton, Terry, Fredric Jameson, and Edward W. Said. 1997. *Nationalism, Colonialism, and Literature.* Minneapolis: University of Minnesota Press.

Easterling, Keller. 2014. *Extrastatecraft: The Power of Infrastructure Space.* Brooklyn: Verso.

Edwards, Paul. 2010. *A Vast Machine: Computer Models, Climate Data, and the Politics of Global Warming.* Cambridge, MA: MIT Press.

E-Flux. 2018. "Project Arts Centre." May 15. https://www.e-flux.com/announcements/198070/jesse-jonestremble-tremble/.

Enterprise Ireland. 2017. *The Future Data Centre: Trends in Smart Data Centre Design and Construction.* Dublin: Enterprise Ireland.

EnterpriseIrelandTV. 2017. "The Irish Advantage." YouTube, September 5. https://www.youtube.com/watch?v=jcjGNAvaCLQ.

Federici, Silvia. 2013. "Permanent Reproductive Crisis: An Interview with Silvia Federici." Interview by Marina Vishmidt. *Mute*, March 7. https://www.metamute.org/editorial/articles/permanent-reproductive-crisis-interview-silvia-federici.

Finn, Christina. 2017. "How Ireland Wants to Become the Data Centre Capital of the World." *TheJournal.ie*, October 22. http://www.thejournal.ie/ireland-data-centres-3656728-Oct2017/.

Fís Éireann/Screen Ireland. 2020. "Fís Éireann/Screen Ireland." Accessed July 29, 2020. https://www.screenireland.ie.

Fitzgerald, Louise. 2023. "What's the Impact of Large-Scale Forestry on Irish Communities?" *RTÉ Brainstorm*, February 30. https://www.rte.ie/brainstorm/2023/0201/1352939-ireland-forestry-plantations-coillte-sitka-spruce-local-communities/.

Florida, Richard. 2005. *Cities and the Creative Class.* New York: Routledge.

Florida, Richard. 2013. "More Losers Than Winners in America's New Economic Geography." *Citylab*, January 30. https://www.citylab.com/life/2013/01/more-losers-winners-americas-new-economic-geography/4465/.

Flynn, Peter. 2011. *Blazing the Trail: The O'Kalems in Ireland.* New York: BIFF Productions.

Furuhata, Yuriko. 2022. *Climatic Media: Transpacific Experiments in Atmospheric Control.* Durham, NC: Duke University Press.

Gabrys, Jennifer. 2016. *Program Earth: Environmental Sensing Technology and the Making of a Computational Planet*. Minneapolis: University of Minnesota Press.

Gago, Verónica. 2017. *Neoliberalism from Below: Popular Pragmatics and Baroque Economics*. Durham, NC: Duke University Press.

Gago, Verónica, and Sandro Mezzadra. 2017. "A Critique of the Extractive Operations of Capital: Toward an Expanded Concept of Extractivism." *Rethinking Marxism* 29, no. 4: 574–91.

Gaiety School of Acting. 2009. "Changes to Section 481 Boost Irish Film Industry." Accessed July 29, 2020. https://gaietyschoolofacting.wordpress.com/2009/05/05/changes-to-section-4/.

Gallagher, Alanna. 2017. "Smart Irishtown Renovations Could Have Google Appeal." *Irish Times*, June 13. https://www.irishtimes.com/life-and-style/homes-and-property/new-to-market/smart-irishtown-renovations-could-have-google-appeal-1.3117995.

Gavin, Tommy. 2020. "Support Apollo House—Framing Vacancy from the Grassroots." Conference presentation from Housing and Ideology: A Symposium Examining the Cultural Impact on an Obedient City, February 24, Dublin.

Gibbons, Luke. 1996. *Transformations in Irish Culture*. South Bend, IN: University of Notre Dame Press.

Gill, Rosalind, and Andy C. Pratt. 2008. "In the Social Factory? Immaterial Labour, Precariousness and Cultural Work." *Theory, Culture and Society* 25, no. 7–8: 1–30.

Gill, Rosalind, Andy C. Pratt, and Tarek E. Virani, eds. 2019. *Creative Hubs in Question: Place, Space, and Work in the Creative Economy*. New York: Palgrave Macmillan.

Gilmore, Ruth Wilson. 2020. "Ruth Wilson Gilmore Makes the Case for Abolition." Interview by Chenjerai Kumanyika. *The Intercept*, June 10. https://theintercept.com/2020/06/10/ruth-wilson-gilmore-makes-the-case-for-abolition/.

Gleeson, Colin. 2018. "Google Buys Bolands Quay Site in Dublin's Docklands." *Irish Times*, May 18. https://www.irishtimes.com/business/technology/google-buys-bolands-quay-site-in-dublin-s-docklands-1.3500256.

Gómez-Barris, Macarena. 2017. *The Extractive Zone: Social Ecologies and Decolonial Perspectives*. Durham, NC: Duke University Press.

Goodbody, Will. 2018. "Inside Facebook's Clonee Data Centre." RTÉ, September 15. https://www.rte.ie/news/ireland/2018/0914/993768-facebook-data-centre/.

Goodbody, Will. 2020. "IDA Ireland Warns of 'Challenging' FDI Outlook amid COVID-19." RTÉ, July 8. https://www.rte.ie/news/business/2020/0708/1152035-ida-ireland-on-fdi/.

Government of Ireland. 2017. *An Bille um Pleanáil agus Forbairt (Bonneagar Straitéiseach) (Leasú)/2017 Planning and Development (Strategic Infra-*

structure) (*Amendment*) *Bill 2017*. Dublin: Stationery Office. https://data
.oireachtas.ie/ie/oireachtas/bill/2017/120/eng/initiated/b12017d.pdf.

Government of Ireland. 2018a. *Government Statement on the Role of Data Centres
in Ireland's Enterprise Strategy*. June. https://dbei.gov.ie/en/Publications
/Publication-files/Government-Statement-Data-Centres-Enterprise
-Strategy.pdf.

Government of Ireland. 2018b. *Project Ireland 2040 National Planning Framework*. https://assets.gov.ie/7338/31f2c0e4ba744fd290206ac0da35f747.pdf.

Graham, Stephen, and Simon Marvin. 2001. *Splintering Urbanism: Networked
Infrastructures, Technological Mobilities and the Urban Condition*. New York:
Routledge.

Graham, Stephen, and Nigel Thrift. 2007. "Out of Order: Understanding Repair
and Maintenance." *Theory, Culture and Society* 24, no. 3: 1–25.

Greenberg, Miriam. 2004. *Branding New York: How a City in Crisis Was Sold to
the World*. New York: Routledge.

Grieveson, Lee. 2017. *Cinema and the Wealth of Nations: Media, Capital, and the
Liberal World System*. Oakland: University of California Press.

Guinan, Kerry. 2016. *The Impact and Instrumentalisation of Art in the Dublin Property Market: Evidence from Smithfield Dublin 1996–2016*. Self-
published. https://www.academia.edu/30498471/The_Impact_and
_Instrumentalisation_of_Art_in_the_Dublin_Property_Market.

Guinan, Kerry (@KerGuinan). 2020. "I'll Start—Once Got Told by DCC
Arts Office That a Court Case Is Not an Artwork and Doesn't Qualify for
Arts Funding." *Twitter*, April 24. https://twitter.com/kerguinan/status
/1253779330725011462?s=21.

Haiven, Max. 2014. *Cultures of Financialization: Fictitious Capital in Popular Culture and Everyday Life*. New York: Palgrave Macmillan.

Halberstam, Jack. 2005. *In a Queer Time and Place: Transgender Bodies, Subcul-
tural Lives*. New York: NYU Press.

Halberstam, Jack. 2020. *Wild Things: The Disorder of Desire*. Durham, NC: Duke
University Press.

Halberstam, Jack, and Tavia Nyong'o. 2018. "Introduction: Theory in the Wild."
South Atlantic Quarterly 117, no. 3: 453–64.

Hall, Stuart. 2007. "Living with Difference: Stuart Hall in Conversation with
Bill Schwarz." Interview by Bill Schwarz. *Soundings: A Journal of Politics
and Culture* 37: 148–58.

Hamilton, Peter. 2017. "JP Morgan Move Cements Capital Dock as Flagship
Development." *Irish Times*, May 15. https://www.irishtimes.com/business
/construction/jp-morgan-move-cements-capital-dock-as-flagship
-development-1.3083448.

Hamilton, Peter. 2019. "Troy Studios Now Largest Production Facility in Ireland
After Stage Expansion." *Irish Times*, December 17. https://www.irishtimes

.com/business/media-and-marketing/troy-studios-now-largest-production
-facility-in-ireland-after-stage-expansion-1.4118293.

Hamilton, Peter. 2020. "South Dublin Council Sells Land for Film Studios for
€26.4m." *Irish Times*, May 12. https://www.irishtimes.com/business/media
-and-marketing/south-dublin-council-sells-land-for-film-studios-for-26
-4m-1.4251388.

Hancock, Ciarán. 2017. "NAMA to 'Resolve' Remaining Ghost Estates by End
of 2017." *Irish Times*, January 5. https://www.irishtimes.com/business
/financial-services/nama-to-resolve-remaining-ghost-estates-by-end-of
-2017-1.2926635.

Haraway, Donna. 2015. "Anthropocene, Capitalocene, Plantationocene, Chthu-
lucene: Making Kin." *Environmental Humanities* 6: 159–65.

Hardt, Michael, and Antonio Negri. 2000. *Empire*. Cambridge, MA: Harvard
University Press.

Harney, Stefano, and Fred Moten. 2013. *The Undercommons: Fugitive Planning
and Black Study*. New York: Minor Compositions.

Harvey, David. 1990. *The Condition of Postmodernity: An Enquiry into the Origins
of Cultural Change*. Malden, MA: Blackwell.

Harvey, David. 2001. "Globalization and the Spatial Fix." *geographische revue* 2:
23–30.

Harvey, David. 2005. *A Brief History of Neoliberalism*. Oxford: Oxford University Press.

Haughey, Anthony. 2018. "A Landscape of Crisis: Photographing Post-Celtic
Tiger Ghost Estates." In *Imagining Suburbia in Literature and Culture*, ed-
ited by Eoghan Smith and Simon Workman, 301–22. New York: Palgrave
Macmillan.

Hearne, Rory. 2013. "Politics and Protest in Ireland: A Brief History and a Call to Action."
Ireland after NAMA, October 14. https://irelandafternama.wordpress.com/2013
/10/14/politics-and-protest-in- ireland-a-brief-history-and-a-call-to-action/.

Hearne, Rory. 2014. "Actually Existing Neoliberalism: Public-Private Partner-
ships in Public Service and Infrastructure Provision in Ireland." In *Neoliberal
Urban Policy and the Transformation of the City: Reshaping Dublin*, edited by
Andrew MacLaran and Sinéad Kelly, 157–73. New York: Palgrave Macmillan.

Hesmondhalgh, David. 2021. "The Infrastructural Turn in Media and Internet
Research." *The Routledge Companion to Media Industries*, edited by Paul
McDonald, 132-142. New York: Routledge.

Hesmondhalgh, David, and Andy C. Pratt. 2005. "Cultural Industries and Cultural Policy." *International Journal of Cultural Policy* 11, no. 1: 1–14.

Higgins, Michael. 2012. "The Poorhouse Revisited." *Vimeo*. http://www
.experimentalfilmsociety.com/film/the-poorhouse-revisited/.

Higgins, Michael D. 2015. "Speech at the 2015 Aosdána General Assembly."
March 5. Keynote address, Royal Hospital, Kilmainham, Dublin, Ireland.

http://www.president.ie/en/media-library/speeches/speech-at-the-2015
-aosdana-general-assembly.

Ho, Karen. 2009. "Disciplining Investment Bankers, Disciplining the Economy:
Wall Street's Institutional Culture of Crisis and the Downsizing of 'Corporate America.'" *American Anthropologist* 111, no. 2: 177–89.

Hockenberry, Matthew, Nicole Starosielski, and Susan Zieger. 2021. *Assembly
Codes: The Logistics of Media*. Durham, NC: Duke University Press.

Hogan, Mél. 2015a. "Data Flows and Water Woes: The Utah Data Center." *Big
Data and Society* 2, no. 2: 1–12.

Hogan, Mél. 2015b. "Facebook Data Storage Centers as the Archive's Underbelly."
Television and New Media 16, no. 1: 3–18.

Hogan, Mél. 2018a. "Big Data Ecologies." *Ephemera: Theory and Politics in
Organization* 18, no. 3: 631–57.

Hogan, Mél. 2018b. "Templating Life: DNA as Nature's Hard Drive." *Public* 57:
145–53.

Hogan, Mél. 2021. "The Data Center Industrial Complex." In *Saturation: An
Elemental Politics*, edited by Melody Jue and Rafico Ruiz. Durham, NC:
Duke University Press.

Holt, Jennifer, and Alisa Perren. 2009. "Introduction: Does the World Really
Need One More Field of Study?" In *Media Industries: History, Theory,
Method*, edited by Jennifer Holt and Alisa Perren, 1–16. Malden, MA:
Wiley-Blackwell.

Holt, Jennifer, and Patrick Vonderau. 2015. "Where the Internet Lives: Data
Centers as Cloud Infrastructures." In *Signal Traffic: Critical Studies of Media
Infrastructures*, edited by Lisa Parks and Nicole Starosielski, 71–93. Cham-
paign: University of Illinois Press.

Honohan, Patrick. 2021. "Is Ireland Really the Most Prosperous Country in
Europe?" *Central Bank of Ireland Economic Letter* 1, February. https://www
.centralbank.ie/docs/default-source/publications/economic-letters/vol
-2021-no-1-is-ireland-really-the-most-prosperous-country-in-europe.pdf
?sfvrsn=25.

Horgan-Jones, Jack. 2023. "Warnings Sounded as Corporation Tax Dips Days
Before Budget Announcement." *Irish Times*, October 4. https://www
.irishtimes.com/politics/2023/10/04/warnings-sounded-as-corporation
-tax-dips-days-before-budget-announcement/.

Houses of the Oireachtas. 2023. *Committee on Budgetary Oversight Report on
Section 481—Film Tax Credit*. May. https://data.oireachtas.ie/ie/oireachtas
/committee/dail/33/committee_on_budgetary_oversight/reports/2023
/2023-05-09_report-on-section-481-film-tax-credit_en.pdf.

Howe, Cymene. 2019. *Ecologics: Wind and Power in the Anthropocene*. Durham,
NC: Duke University Press.

Hulme, Mike. 2017. *Weathered: Cultures of Climate*. New York: Sage.

Humphries, Conor, and Lois Kapila. 2019. "A City Tour Reveals the Infrastructure of the Internet All Around Us." *Dublin Inquirer*, May 22. https://dublininquirer.com/2019/05/22/a-city-tour-reveals-the-infrastructure-of-the-internet-all-around-us.

IDA Ireland. 2018. *A Study of the Economic Benefits of Data Centre Investment in Ireland*. May. https://www.idaireland.com/newsroom/publications/ida-ireland-economic-benefits-of-data-centre-inves.

IFTN. 2017. "Dublin Bay Studios Promoters Alan Moloney and James Morris Endorsed by Dublin City Council for 'Visionary' Project." May 19. http://www.iftn.ie/news/?act1=record&only=1&aid=73&rid=4290665&tpl=archnews&force=1.

Ignatiev, Noel. 1995. *How the Irish Became White*. New York: Routledge.

Image Now. n.d. "Bolands Quay Designing Dublin's Docklands." http://www.imagenow.ie/projects/bolands-quay.

Irish Environment. 2012. "Ghost Estates." May 30. http://www.irishenvironment.com/iepedia/ghost-estates/.

Irish Film Board. 2016. *Strategic Plan 2016–2020: Building on Success*. https://www.screenireland.ie/images/uploads/general/IFB_Five_Year_Strategy_2016-1.pdf.

Irish Glass Bottle Housing Action Group. 2019. "DATE FOR YOUR DIARY!!!" *Facebook*, May 16. https://www.facebook.com/IGBhousing/photos/a.1607147112909870/2159125197712056/.

Irish History Podcast (@irishhistory). 2019. "If Notre Dame Was in Dublin, There Would Be Surveyors in the Ruins Right Now Figuring How They Could Convert It into a Hotel or a Hub for Tax Exempt Tech Startups." *Twitter*, April 16. https://twitter.com/irishhistory/status/1118071399170617344.

Irish Times. 2016. "Over 2,000 March in Athenry in Support of Apple Data Centre." *Irish Times*, November 6. https://www.irishtimes.com/business/technology/over-2-000-march-in-athenry-in-support-of-apple-data-centre-1.2856916.

IWEA. 2020. *Data-Centre Implications for Energy Use in Ireland*. https://windenergyireland.com/images/files/9660bdc9488d1ce95c4f2ba74ce34obfb4831f.pdf.

Jackson, Steve J. 2014. "Rethinking Repair." In *Media Technologies: Essays on Communication, Materiality and Society*, edited by Tarleton Gillespie, Pablo Boczkowski, and Kirsten Foot. Cambridge, MA: MIT Press.

Jacobson, Brian. 2015. *Studios Before the System: Architecture, Technology, and the Emergence of Cinematic Space*. New York: Columbia University Press.

Jain, Isha. 2024. "Data Centres in Ireland to Consume One-Third of Electricity by 2026: Schneider Electric Responds." *Data Centre Network and News*, January 26.

James, Kevin J. 2014. *Tourism, Land and Landscape in Ireland: The Commodification of Culture*. New York: Routledge.

Jenkins, Henry. 2006. *Convergence Culture: Where Old and New Media Collide.* New York: NYU Press.

Johnson, Alix. 2019. "Emplacing Data Within Imperial Histories: Imagining Iceland as Data Centers' 'Natural' Home." *Culture Machine* 18. https://culturemachine.net/vol-18-the-nature-of-data-centers/emplacing-data/.

Jones, Campbell. 2020. "Introduction: The Return of Economic Planning." *South Atlantic Quarterly* 119, no. 1: 1–10.

Jones, Huw David. 2016. "The Cultural and Economic Implications of UK/European Co-Production." *Transnational Cinemas* 7, no. 1: 1–20.

Jones, Penny. 2014. "Discovering Ireland." *Data Center Dynamics*, January 6. http://www.datacenterdynamics.com/content-tracks/power-cooling/discovering-ireland/84470.fullarticle.

Judge, Peter. 2019. "Floating Data Center Approved for Launch in Ireland." *Data Center Dynamics*, September 6. https://www.datacenterdynamics.com/en/news/floating-data-center-approved-launch-ireland/.

Kearney, Richard. 1996. *Postnationalist Ireland: Politics, Culture, Philosophy.* London: Routledge.

Kelly, Olivia. 2016. "Up to 3,000 homes to be built on former Irish Glass Bottle site." *Irish Times*, 16 December. https://www.irishtimes.com/news/environment/up-to-3-000-homes-to-be-built-on-former-irish-glass-bottle-site-1.2907397.

Kelly, Olivia. 2018. "Pigeon House Power Station and Hotel Set for Redevelopment." *Irish Times*, May 11. https://www.irishtimes.com/culture/heritage/pigeon-house-power-station-and-hotel-set-for-redevelopment-1.3492202.

Kelly, Olivia. 2019. "Five Companies Competing to Redevelop Dublin's Pigeon House Site." *Irish Times*, January 7. https://www.irishtimes.com/culture/heritage/five-companies-competing-to-redevelop-dublin-s-pigeon-house-site-1.3749361.

Kelly, Sinéad. 2014. "Light-Touch Regulation: The Rise and Fall of the Irish Banking Sector." In *Neoliberal Urban Policy and the Transformation of the City: Reshaping Dublin*, edited by Andrew MacLaran and Sinéad Kelly, 37–52. New York: Palgrave Macmillan.

Kennedy, Sinéad. 2003. "Irish Women and the Celtic Tiger Economy." In *The End of Irish History? Reflections on the Celtic Tiger*, edited by Colin Coulter and Steve Coleman, 95–109. Manchester: Manchester University Press.

Kennedy, Wilson. 2016. "Capital Dock." YouTube, September 20. https://www.youtube.com/watch?v=hs4DWEMpdSg.

Kerr, Aphra. 2000. "Media Diversity and Cultural Identities: The Development of Multimedia 'Content' in Ireland." *New Media and Society* 2, no. 3: 286–312.

Kerr, Aphra. 2014. "Placing International Media Production." *Media Industries* 1, no. 1: 27–32.

Kerr, Aphra, and Paschal Preston. 2001. "Digital Media, Nation-States, and Local Cultures: The Case of Multimedia 'Content' Production." *Media, Culture, and Society* 23, no. 1: 109–31.

Kerrigan, Páraic, and Anne O'Brien. 2020. "Camping It Up and Toning It Down: Gay and Lesbian Sexual Identity in Media Work." *Media, Culture and Society* 42, nos. 7–8: 1068–77. https://doi.org/10.1177%2F0163443720908149.

Kidman, Shawna. 2019. "Introduction." In *Comic Books Incorporated: How the Business of Comics Became the Business of Hollywood*, 1–17. Berkeley: University of California Press.

Kilraine, John. 2019. "Permission Granted for 3,500 New Homes at Dublin's Poolbeg." RTÉ, April 11. https://www.rte.ie/news/dublin/2019/0411/1042035 -poolbeg-housing/.

Kitchin, Rob, Cian O'Callaghan, Mark Boyle, Justin Gleeson, and Karen Keaveney. 2012. "Placing Neoliberalism: The Rise and Fall of Ireland's Celtic Tiger." *Environment and Planning A: Economy and Space* 44, no. 6: 1302–26.

Kitchin, Rob, Justin Gleeson, Karen Keaveney, and Cian O'Callaghan. 2010. "A Haunted Landscape: Housing and Ghost Estates in Post-Celtic Tiger Ireland." *NIRSA Working Paper Series* 59: 1–66.

Kitching, Karl. 2015. "How the Irish Became CRT'd? 'Greening' Critical Race Theory, and the Pitfalls of a Normative Atlantic State View." *Race, Ethnicity and Education* 18, no. 2: 163–82.

Klein, Naomi. 2007. *The Shock Doctrine: The Rise of Disaster Capitalism*. New York: Picador.

Laird, Heather. 2015. "European Postcolonial Studies and Ireland: Towards a Conversation Amongst the Colonized of Europe." *Postcolonial Studies* 18, no. 4: 384–96.

Lally, Donal. 2021. "The Cybernetic Wilderness of Data Centres." In *States of Entanglement: Data in the Irish Landscape*, edited by ANNEX. New York: Actar.

Larkin, Brian. 2008. *Signal and Noise: Media, Infrastructure, and Urban Culture in Nigeria*. Durham, NC: Duke University Press.

Larkin, Brian. 2013. "The Politics and Poetics of Infrastructure." *Annual Review of Anthropology* 42: 327–43.

Lash, Scott, and John Urry. 1987. *The End of Organized Capitalism*. Madison: University of Wisconsin Press.

Lawton, Phil. 2013. "Rethinking the Livable City in a Post Boom-Time Ireland." In *Spacing Ireland: Place, Society and Culture in a Post-Boom Era*, edited by Caroline Crowley and Denis Linehan, 102–15. Manchester: Manchester University Press.

Lawton, Phil, Enda Murphy, and Declan Redmond. 2014. "Neoliberalising the City 'Creative-Class' Style." In *Neoliberal Urban Policy and the Transformation of the City: Reshaping Dublin*, edited by Andrew MacLaran and Sinéad Kelly, 189–202. New York: Palgrave Macmillan.

Lazzarato, Maurizio. 1996. "Immaterial Labor." In *Radical Thought in Italy: A Potential Politics*, translated by Paul Colilli and Ed Emory, edited by Paolo Virno and Michael Hardt, 142–57. Minneapolis: University of Minnesota Press.

Lazzarato, Maurizio. 2012. *The Making of Indebted Man: An Essay on the Neoliberal Condition*. Translated by Joshua David Jordan. Los Angeles: Semiotext(e).

LeCavalier, Jesse. 2016. *The Rule of Logistics: Walmart and the Architecture of Fulfillment*. Minneapolis: University of Minnesota Press.

Lee, Benjamin, and Edward LiPuma. 2002. "Cultures of Circulation: Imaginations of Modernity." *Public Culture* 14, no. 1: 191–213.

Lee, Benjamin, and Edward LiPuma. 2004. *Financial Derivatives and the Globalization of Risk*. Durham, NC: Duke University Press.

Lefebvre, Henri. 1991 (1974). *The Production of Space*. Translated by Donald Nicholson Smith. Cambridge, MA: Blackwell.

Levidow, Les. 2002. "Marketizing Higher Education: Neoliberal Strategies and Counter-Strategies." In *The Virtual University? Knowledge, Markets and Management*, edited by Kevin Robins and Frank Webster, 227–48. Oxford: Oxford University Press.

Levinson, Marc. 2016. *The Box: How the Shipping Container Made the World Smaller and the World Economy Bigger*. Princeton, NJ: Princeton University Press.

Liang, Lawrence. 2005. "Porous Legalities and Avenues of Participation." In *Sarai Reader 05: Bare Acts*, edited by Lawrence Liang, 6–17. Delhi: Sarai Programme.

Lima, João Marques. 2019. "Ireland Needs Urgent $10bn Energy Investment to Power Data Centres." *Data Economy*, August 23. https://data-economy.com/ireland-needs-urgent-10bn-energy-investment-to-power-data-centres/.

Limerick.ie. n.d. "Creative Limerick." Accessed July 29, 2020. https://www.limerick.ie/creative-limerick.

Linehan, Denis, and Caroline Crowley. 2013. "Introduction: Geographies of the Post-Boom Era." In *Spacing Ireland: Place, Society and Culture in a Post-Boom Era*, edited by Caroline Crowley and Denis Linehan, 1–14. Manchester: Manchester University Press.

LiPuma, Edward. 2017. *The Social Life of Financial Derivatives: Markets, Risk, and Time*. Durham, NC: Duke University Press.

Lloyd, David. 2003. "Rethinking National Marxism: James Connolly and 'Celtic Communism.'" *Interventions: International Journal of Postcolonial Studies* 5, no. 3: 345–70.

Lobato, Ramon. 2012. *Shadow Economies of Cinema: Mapping Informal Film Distribution*. New York: Palgrave Macmillan.

Lovink, Geert, and Ned Rossiter. 2007. "Proposals for Creative Research: Introduction to the *MyCreativity Reader*." In *MyCreativity Reader: A Critique of Creative Industries*, edited by Geert Lovink and Ned Rossiter, 11–17. Amsterdam: Institute of Network Cultures.

Lynch, Suzanne. 2017. "Apple Fails to Confirm Athenry Data Centre Plans at Varadkar Meeting." *Irish Times*, November 3. https://www.irishtimes.com /business/technology/apple-fails-to-confirm-athenry-data-centre-plans-at -varadkar-meeting-1.3279061.

MacArthur, Joan. 2017. "Poolbeg West—Draft Planning Scheme." SAMRA, July 3. https://samra.ie/wp-content/uploads/2018/12/P20170703-SAMRA -POOLBEG-SDZ-Rev-2-3-7-17-Letter-Dublin-Docklands-DCC.pdf.

MacLaran, Andrew, and Sinéad Kelly. 2014a. "Irish Neoliberalism and Neoliberal Urban Policy." In *Neoliberal Urban Policy and the Transformation of the City: Reshaping Dublin*, edited by Andrew MacLaran and Sinéad Kelly, 20–36. New York: Palgrave Macmillan.

MacLaran, Andrew, and Sinéad Kelly, eds. 2014b. *Neoliberal Urban Policy and the Transformation of the City: Reshaping Dublin.* New York: Palgrave Macmillan.

Marazzi, Christian. 2011. *The Violence of Financial Capitalism.* Translated by Kristina Lebedeva and Jason Francis McGimsey. Los Angeles: Semiotext(e).

Marks, Laura. 2020. "Streaming Video, a Link Between Pandemic and Climate Crisis." *Harun Farocki Institut*, April 16. https://www.harun-farocki-institut .org/en/2020/04/16/streaming-video-a-link-between-pandemic-and -climate-crisis-journal-of-visual-culture-hafi-2/?fbclid=IwAR3jb2gG4Xye8f -ou5NijLPBPJaA-6axLcbKDk4Jb_BeihTD23AlBqRD9Qw.

Marx, Karl. 1992. *Capital*, vol. 1: *A Critique of Political Economy.* Translated by Ben Fowkes. New York: Penguin Classics.

Massey, Doreen. 2005. *For Space.* New York: Sage.

Mattern, Shannon. 2013. "Infrastructural Tourism." *Places*, July. https:// placesjournal.org/article/infrastructural-tourism/?cn-reloaded=1.

Mattern, Shannon. 2017. *Code, Clay, Data, and Dirt: Five Thousand Years of Urban Media.* Minneapolis: University of Minnesota Press.

Mattern, Shannon. 2018. "Maintenance and Care." *Places*, November. https:// placesjournal.org/article/maintenance-and-care/.

Maxwell, Richard. 2014. "Media Industries and the Ecological Crisis." *Media Industries Journal* 1, no. 2: 36–39.

Mayer, Vicki. 2011. *Below the Line: Producers and Production Studies in the New Television Economy.* Durham, NC: Duke University Press.

Mayer, Vicki. 2017. *Almost Hollywood, Nearly New Orleans: The Lure of the Local Film Economy.* Berkeley: University of California Press.

Mayer, Vicki. 2023. "When Do We Go from Here? Data Center Infrastructure Labor, Jobs, and Work in Economic Development Time and Temporalities." *New Media and Society* 25, no. 2: 307–23.

Mayer, Vicki, and Julia Velkova. 2023. "This Site Is a Dead End? Employment Uncertainties and Labor in Data Centers." *Information Society*, vol. 39, no. 2: 112–22.

Mayer, Vicki, Miranda J. Banks, and John T. Caldwell. 2009. "Introduction: Production Studies: Routes and Routes." In *Production Studies: Cultural Studies of Media Industries*, 1–12. New York: Routledge.

McCabe, Conor. 2022. "Apple and Ireland, 1980–2020: A Case Study of the Irish Comprador Capitalist System." *Radical History Review* 143: 141–48.

McCabe, Conor (@CMacCaba). 2020. "There's a Creation Myth About Modern Ireland That Needs Debunking." *Twitter*, August 11. https://twitter.com/CMacCaba/status/1293329614924914688.

McCarthy, Anna. 2021. "The Angelus: Devotional Television, Changing Times." *Television and New Media* 22, no. 1: 12–31.

McCormack, Derek. 2018. *Atmospheric Things: On the Allure of Environmental Envelopment*. Durham, NC: Duke University Press.

McDonald, Henry. 2012. "Ireland Is Cool for Google as Its Data Servers Like the Weather." *The Guardian*, December 23. https://www.theguardian.com/technology/2012/dec/23/ireland- cool-google-data-servers-weather.

McGovern, Mark. 2002. "'The 'Craic' Market': Irish Theme Bars and the Commodification of Irishness in Contemporary Britain." *Irish Journal of Sociology* 11, no. 2: 77–98.

McLaughlin, John. 2015. "Data: Clouds and Precipitation." In *Infrastructure and the Architectures of Modernity in Ireland 1916–2016*, edited by Gary A. Boyd and John McLaughlin, 185–99. Burlington, VT: Ashgate.

McLoone, Martin. 2000. *Irish Film: The Emergence of a Contemporary Cinema*. London: British Film Institute.

McLoone, Martin. 2008. *Film, Media and Popular Culture in Ireland: Cityscapes, Landscapes, Soundscapes*. Dublin: Irish Academic Press.

McMahon, Páraic. 2019. "Ennis Data Centre 'Will Be the Same as Having a Goldmine.'" *Clare Echo*, March 23. https://www.clareecho.ie/ennis-data-centre-will-be-the-same-as-having-a-goldmine/.

McVeigh, Robbie, and Bill Rolston. 2021. *Ireland, Colonialism, and the Unfinished Revolution: Anois Ar Theacht an TSamhraidh*. Dublin: Beyond the Pale.

McWilliams, David. 2006. "A Warning from Deserted Ghost Estates." *David McWilliams*, October 1. http://www.davidmcwilliams.ie/a-warning-from-deserted-ghost-estates/.

Meade, Melissa. 2017. "In the Shadow of the Coal Breaker: *Cultural Extraction* and Participatory Communication in the Anthracite Mining Region." *Cultural Studies* 31, nos. 2–3: 376–99.

Melhuish, Clare, Monica Degen, and Gillian Rose. 2016. "'The Real Modernity That Is Here': Understanding the Role of Digital Visualisations in the Production of a New Urban Imaginary at Msheireb Downtown, Doha." *City and Society* 28, no. 2: 222–45.

Melia, Paul. 2013. "EU Court Blocks Bypass over Fears of Damage to Protected Limestone." *Independent.ie*, April 12. https://www.independent.ie

/irish-news/eu-court-blocks-bypass-over-fears-of-damage-to-protected-limestone-29191203.html.

Mercier, Sinéad. 2021. "Ireland's Energy System: The Historical Case for Hope in Climate Action." *New Labor Forum* 30, no. 2: 21–30.

Mercille, Julien. 2014. "The Role of the Media in Sustaining Ireland's Housing Bubble." *New Political Economy* 19, no. 2: 282–301.

Mezzadra, Sandro. 2013. "Double Opening, Split Temporality, and New Spatialities: An Interview with Sandro Mezzadra on 'Militant Research.'" Interview by Glenda Garelli and Martina Tazzioli. *Postcolonial Studies* 16, no. 3: 309–19.

Mezzadra, Sandro, and Brett Neilson. 2013a. *Border as Method: Or, the Multiplication of Labor*. Durham, NC: Duke University Press.

Mezzadra, Sandro, and Brett Neilson. 2013b. "Extraction, Logistics, Finance: Global Crisis and the Politics of Operations." *Radical Philosophy* 178: 8–18.

Mezzadra, Sandro, and Brett Neilson. 2015. "Operations of Capital." *South Atlantic Quarterly* 114, no. 1: 1–9.

Mezzadra, Sandro, and Brett Neilson. 2017. "On the Multiple Frontiers of Extraction: Excavating Contemporary Capitalism." *Cultural Studies* 31, nos. 2–3: 185–204.

Mezzadra, Sandro, and Brett Neilson. 2019. *The Politics of Operations: Excavating Contemporary Capitalism*. Durham, NC: Duke University Press.

Miller, Toby, Richard Maxwell, Nitin Govil, and John McMurria. 2001. *Global Hollywood*. London: British Film Institute.

Mitchell, Timothy. 2011. *Carbon Democracy: Political Power in the Age of Oil*. Brooklyn: Verso.

Moore, Hayden. 2020. "Excess Amazon Heat to Give Low-Cost Hot Water to Tallaght Area." *Echo.ie*, February 19. https://www.echo.ie/show/article/excess-amazon-heat-to-give-low-cost-hot-water-to-tallaght-area.

Moore, Jason W. 2011. "Wall Street Is a Way of Organizing Nature." Interview by Tom Keefer. *Upping the Anti: A Journal of Theory and Action* 12: 39–53.

Moore, Jason W. 2017. "The Capitalocene, Part I: On the Nature and Origins of Our Ecological Crisis." *Journal of Peasant Studies* 44, no. 3: 594–630.

Morash, Christopher. 2010. *A History of the Media in Ireland*. Cambridge: Cambridge University Press.

Mulhall, Anne. 2013. "The Feng Shui of Lough Derg: Therapeutic Landscapes and the Marketing of Spirituality in Contemporary Ireland." In *Viewpoints: Theoretical Perspectives on Irish Visual Texts*, edited by Claire Bracken and Emma Radley, 13–28. Cork: Cork University Press.

Mulhall, Anne. 2020. "The Ends of Irish Studies? On Whiteness, Academia, and Activism." *Irish University Review* 50, no. 1: 94–111.

Mullally, Una. 2017. "Why Are Stunning Dublin Murals Being Painted Over?" *Irish Times*, December 2. https://www.irishtimes.com/culture/art-and-design/why-are-stunning-dublin-murals-being-painted-over-1.3309756.

Mullally, Una. 2018. "Artist Group Subset Calls on Council to Change Street Art Regulations." *Irish Times*, June 15. https://www.irishtimes.com/culture /art-and-design/visual-art/artist-group-subset-calls-on-council-to-change -street-art-regulations-1.3531933.

Murphy, Denis. 2018. "Ardmore Studios, Film Workers, and the Irish State: Creative Labour in the 'Decade of Upheaval.'" *Saothar: Journal of the Irish Labour History Society* 43: 69–82.

Murphy, Denis, and Maria O'Brien. 2015. "Irish Film Finance Rebooted: The New Section 481." *Estudios Irlandeses* 10: 225–27.

Murphy, Enda, Linda Fox-Rogers, and Berna Grist. 2014. "The Political Economy of Legislative Change: Neoliberalising Planning Legislation." In *Neoliberal Urban Policy and the Transformation of the City: Reshaping Dublin*, edited by Andrew MacLaran and Sinéad Kelly, 53–65. New York: Palgrave Macmillan.

Murphy, Niall. 2018. "#Industry: Olcott Entertainment Strikes Deal for Ardmore Studios." *Scannain: Irish for Movies*, March 23. https://scannain.com/irish /industry/ardmore-studios-olcott-entertainment-purchase/.

National Asset Management Agency. n.d. "Our Work." Accessed July 29, 2020. https://www.nama.ie/our-work.

Neilson, Brett, and Ned Rossiter. 2008. "Precarity as a Political Concept, or, Fordism as Exception." *Theory, Culture and Society* 25, nos. 7–8: 51–72.

Neilson, Brett, and Ned Rossiter. 2017. "'Buy Cheap, Sell Dear.'" In *Logistical Worlds: Infrastructure, Software, Labor*, edited by Brett Neilson and Ned Rossiter, 7–14. London: Open Humanities Press.

Neves, Joshua. 2013. "For the City Yet to Come: Planning's New Visual Culture." In *Sarai Reader 09: Projections*, edited by Raqs Media Collective and Shveta Sarda, 291–97. http://archive.sarai.net/files/original/afb232aa9790b94453 a01dc2b09cc4aa.pdf.

Newenham, Pamela. 2016. "An Bord Pleanála Gives Green Light for Apple Data Centre." *Irish Times*, August 12. https://www.irishtimes.com/business /construction/an-bord-plean%C3%A1la-gives-green-light-for-apple-data -centre-1.2754006.

Nitzsche, Susan. 2013. "From Periphery to Core (and Back)? Political, Journalistic, and Academic Perceptions of Celtic Tiger- and Post-Celtic Tiger-Ireland." *Studi irlandesi: A Journal of Irish Studies* 3: 115–35.

Nixon, Rob. 2011. *Slow Violence and the Environmentalism of the Poor*. Cambridge, MA: Harvard University Press.

Nye, Joseph S. 1990. "Soft Power." *Foreign Policy* 80: 153–71.

O'Boyle, Brian, and Kieran Allen. 2021. *Tax Haven Ireland*. London: Pluto Press.

O'Brien, A. 2014. "'Men Own Television': Why Women Leave Media Work." *Media, Culture & Society* 36, no. 8: 1207–18. https://doi.org/10.1177/0163443714544868.

O'Brien, Anne, Sarah Arnold, and Páraic Kerrigan. 2021. *Media Graduates at Work: Irish Narratives on Policy, Education and Industry*. London: Palgrave Macmillan.

O'Brien, Anne, Páraic Kerrigan, and Susan Liddy. 2022. "Conceptualising Change in Equality, Diversity and Inclusion: A Case Study of the Irish Film and Television Sector." *European Journal of Cultural Studies* 26, no. 3: 336–53.

O'Callaghan, Cian, Mark Boyle, and Rob Kitchin. 2014. "Post-Politics, Crisis, and Ireland's 'Ghost Estates.'" *Political Geography* 42: 121–33.

O'Callaghan, Cian, and Cesare Di Feliciantonio, eds. 2023. *The New Urban Ruins: Vacancy, Urban Politics and International Experiments in the Post-Crisis City.* Bristol: Bristol University Press.

O'Callaghan, Cian, Cesare Di Feliciantonio, and Michael Byrne. 2018. "Governing Urban Vacancy in Post-Crash Dublin: Contested Property and Alternative Social Projects." *Urban Geography* 39, no. 6: 868–91.

O'Callaghan, Cian, Sinéad Kelly, Mark Boyle, and Rob Kitchin. 2015. "Topologies and Topographies of Ireland's Neoliberal Crisis." *Space and Polity* 19, no. 1: 31–46.

O'Callaghan, Cian, and Kathleen Stokes. 2021. "We Need to Change How We Think About Urban Vacancy." *Dublin Inquirer*, February 24.

O'Connell, Brian, and Cian O'Carroll. 2018. *Brendan O'Regan: Irish Innovator, Visionary and Peacemaker.* Kildare: Irish Academic Press.

O'Connell, Dióg. 2010. *New Irish Storytellers: Narrative Strategies in Film.* Chicago: Intellect.

O'Donoghue, Paul. 2017. "Why Apple Has Been Stuck in a Three-Year Battle over Its Athenry Data Centre." *TheJournal.ie*, September 14. www.thejournal.ie /apple-data-centre-athery-explained-2- 3597680-Sep2017/.

O'Donoghue, Paul. 2018. "Amazon's Grand Plans for Its New Dublin Data Centre Could Use as Much Power as a City." *TheJournal.ie*, January 14. https://www .thejournal.ie/amazon-data-centre-dublin-mulhuddart-3-3805691-Jan2018/.

O'Dwyer, Davin. 2015. "Ireland's Data Centre Boom Set to Continue." *Irish Times*, March 5. https://www.irishtimes.com/business/technology/ireland-s-data -centre-boom-set-to-continue-1.2126081.

O'Hagan, John, Denis Murphy, and Ruth Barton. 2020. "Do State Funding, Geographic Location, and Networks Matter? The Case of Prominent Irish Actors, Directors and Writers." *Cultural Trends* 29, no. 2: 77–95.

O'Halloran, Barry. 2017. "IDA to Seek Advance Planning Permission for Future Data Centres." *Irish Times*, October 16. https://www .irishtimes.com/business/work/ida-to-seek-advance-planning - permission-for-future-data-centres-1.3256935.

O'Halloran, Marie. 2017. "Ardmore Studios Site Planning Restrictions Will Maintain It as Film-Only Zone." *Irish Times*, January 26. https://www.irishtimes .com/news/politics/oireachtas/ardmore-studios-site-planning-restrictions -will-maintain-it-as-film-only-zone-1.2952946

O'Hearn, Denis. 1990. "The Road from Import-Substituting to Export-Led Industrialization in Ireland: Who Mixed the Asphalt, Who Drove the Machinery, and Who Kept Making Them Change Directions?" *Politics and Society,* 18 no. 1: 1–38.

O'Hearn, Denis. 2000. "Globalization, 'New Tigers,' and the End of the Developmental State? The Case of the Celtic Tiger." *Politics and Society* 28, no. 1: 67–92.

O'Hearn, Denis. 2001. *The Atlantic Economy: Britain, the US and Ireland.* Manchester: Manchester University Press.

O'Hearn, Denis. 2016. "Ireland in the World System: An Interview with Denis O'Hearn." Interview by Aidan Beatty, Sharae Deckard, and Maurice Coakley. *Journal of World-Systems Research* 22, no. 1: 202–13.

Ohlmeyer, Jane. 2023. *Making Empire: Ireland, Imperialism, and the Early Modern World.* Oxford: Oxford University Press.

Olsberg SPI. 2023. *The Cultural Dividend Generated by Ireland's Section 481 Film and Television Incentive.* Olsberg SPI and Screen Ireland, January 10. https://www.screenireland.ie/images/uploads/general/The_Cultural_Dividend_Generated_by_Section_481_Report_January_2023.pdf.

Olsberg SPI with Nordicity. 2017. *Economic Analysis of the Audiovisual Sector in the Republic of Ireland.* https://www.chg.gov.ie/app/uploads/2018/06/economic-analysis-of-the-audiovisual-sector-in-the-republic-of-ireland.pdf.

O'Mahoney, Eoin, and Philip Lawton. 2019. "Image Circulation, Development Hoardings and New Strategies of Urban Change." *Geografiska Annaler: Series B, Human Geography* 101, no. 3: 202–18.

O'Neill, Paul. 2018. "On Dublin, Amazon, and the 'Secret Region.'" *Dublin Inquirer,* November 28. https://www.dublininquirer.com/2018/11/28/paul-on-dublin-amazon-and-the-secret-region.

O'Neill, Paul. 2019. "A Guided Tour of Dublin's Physical Internet Infrastructure." Interview by Régine Debatty. *We Make Money Not Art,* September 27. https://we-make-money-not-art.com/a-guided-tour-of-dublins-physical-internet-infrastructure/.

Ong, Aihwa. 2000. *Flexible Citizenship: The Cultural Logics of Transnationality.* Durham, NC: Duke University Press.

Ong, Aihwa. 2006. *Neoliberalism as Exception: Mutations in Citizenship and Sovereignty.* Durham, NC: Duke University Press.

Opaque Media: A Workshop. 2017. Accessed July 29, 2020. https://opaquemediaworkshop.wordpress.com/cfp/.

Ó Riain, Séan. 2000. "The Flexible Developmental State: Globalization, Information Technology and the 'Celtic Tiger.'" *Politics and Society* 28, no. 2: 157–93.

Ormston House. n.d. "About." Accessed July 29, 2020. http://ormstonhouse.com/about/.

O'Sullivan, Kevin. 2019. "US Academic Criticises ESB's Use of Coal from Colombia Open-Pit Mines." *Irish Times,* January 21. https://www.irishtimes

.com/news/environment/us-academic-criticises-esb-s-use-of-coal-from
-colombia-open-pit-mines-1.3765954.

O'Toole, Fintan. 1997. *The Ex-Isle of Erin: Images of a Global Ireland*. Dublin: New Island Books.

O'Toole, Fintan. 2015. "Beyond Belief—Why Did We Grant Disney's Skelligs Wish?" *Irish Times*, September 1. https://www.irishtimes.com/opinion /fintan-o-toole-beyond-belief-why-grant-disney-s-skelligs-wish-for-star -wars-1.2335310.

Parker, Matt. 2019. "An Apple a Day: Listening to Data Centre Site Selection Through a Sonospheric Investigation." *Culture Machine* 18: 1–15. https:// culturemachine.net/vol-18-the-nature-of-data- centers/an-apple-a-day/.

Parker, Matt. n.d. "Fields of Athenry." Accessed July 29, 2020. https://www .earthkeptwarm.com/fields-of-athenry/.

Parks, Lisa. 2005. *Cultures in Orbit: Satellites and the Televisual*. Durham, NC: Duke University Press.

Parks, Lisa. 2009. "Around the Antenna Tree: The Politics of Infrastructural Visibility." *Flow Journal*, March 6. https://www.flowjournal.org/2009/03 /around-the-antenna-tree-the-politics-of-infrastructural-visibilitylisa-parks -uc-santa-barbara/.

Parks, Lisa. 2015a. "'Stuff You Can Kick': Toward a Theory of Media Infrastruc-tures." In *Between Humanities and the Digital*, edited by Patrik Svensson and David Theo Goldberg, 355–73. Cambridge, MA: MIT Press.

Parks, Lisa. 2015b. "Water, Energy, Access: Materializing the Internet in Rural Zam-bia." In *Signal Traffic: Critical Studies of Media Infrastructures*, edited by Lisa Parks and Nicole Starosielski, 115–36. Champaign: University of Illinois Press.

Parks, Lisa, and Caren Kaplan, eds. 2017. *Life in the Age of Drone Warfare*. Durham, NC: Duke University Press.

Parks, Lisa, Lindsay Palmer, and Daniel Grinberg. 2017. "Media Fieldwork: Crit-ical Reflections on Collaborative ICT Research in Rural Zambia." In *Applied Media Studies*, edited by Kirsten Ostherr, 97–116. New York: Routledge.

Parks, Lisa, and Nicole Starosielski, eds. 2015. *Signal Traffic: Critical Studies of Media Infrastructures*. Champaign: University of Illinois Press.

Parks, Lisa, Julia Velkova, and Sander de Ridder, eds. 2023. *Media Backends: Digital Infrastructures and Sociotechnical Relations*. Champaign: University of Illinois Press.

Paul, Mark. 2016. "Apple Remains Stuck in Mud in Fields of Athenry." *Irish Times*, October 28. https://www.irishtimes.com/business/technology/apple -remains-stuck-in-mud-in-fields-of- athenry-1.2845918.

Perren, Alisa. 2013. "Rethinking Distribution for the Future of Media Industry Studies." *Cinema Journal* 52, no. 3: 165–71.

Peters, John Durham. 2015. *The Marvelous Clouds: Toward a Philosophy of Ele-mental Media*. Chicago: University of Chicago Press.

Pettitt, Lance. 2000. *Screening Ireland: Film and Television Representation*. Manchester: Manchester University Press.

Phelan, Sean. 2014. *Neoliberalism, Media, and the Political*. New York: Palgrave Macmillan.

Plantin, Jean-Christophe, and Aswin Punathambekar. 2019. "Digital Media Infrastructures: Pipes, Platforms, and Politics." *Media, Culture and Society* 41, no. 2: 163–74.

Posor, David. 2014. "'End of the Line': Griffith College Student Documents a Dublin Art Space's Last Exhibition." *The Circular*, April 23. http://thecircular.org/end-of-the-line-griffith-college-student-documents-the-last-days-of-a-dublin-art-space/.

Povinelli, Elizabeth. 2011. *Economies of Abandonment: Social Belonging and Endurance in Late Liberalism*. Durham, NC: Duke University Press.

Povinelli, Elizabeth. 2016. *Geontologies: A Requiem to Late Liberalism*. Durham, NC: Duke University Press.

Power, Dominic, and Patrick Collins. 2021. "Peripheral Visions: The Film and Television Industry in Galway, Ireland." *Industry and Innovation* 28, no. 9: 1150–74.

Preston, Paschal, Aphra Kerr, and Anthony Cawley. 2009. "Innovation and Knowledge in the Digital Media Sector: An Information Economy Approach." *Information, Communication, and Society* 7: 994–1014.

PricewaterhouseCoopers. 2016. *Making Sense of a Complex World: Film Financing Arrangements. Media Industry Accounting Group* 11. https://www.pwc.fr/fr/assets/files/pdf/2016/07/film-financing-arrangements.pdf.

PricewaterhouseCoopers. 2020. *Section 481 and the Film/TV Industry: Insights on the Sector's Contribution to the Irish Economy*. https://www.screenireland.ie/images/uploads/general/Screen_Industry_EIA_report.pdf.

radiokerrynews. 2017. "Star Wars May Return to Kerry This Month." *Radio Kerry*, March 3. https://www.radiokerry.ie/star-wars-may-return-west-kerry-month/.

Rampazzo, Giovanna. 2017. "The Production of *Ek Tha Tiger*: A Marriage of Convenience Between Bollywood and the Irish Film and Tourist Industries." *South Asian Popular Culture* 14, no. 3: 167–84.

Ramsey, Phil, Steven Baker, and Robert Porter. 2019. "Screen Production on the 'Biggest Set in the World': Northern Ireland Screen and the Case of *Game of Thrones*." *Media, Culture and Society* 41, no. 6: 845–62.

Ranchordas, Sofia. 2018. "Citizens as Consumers in the Data Economy: The Case of Smart Cities." *Journal of European Consumer and Market Law* 7, no. 4: 154–61.

Reddan, Fiona. 2019. "No Social Housing at Trio of Top Dublin Docklands Developments." *Irish Times*, April 30. https://www.irishtimes.com/business/no-social-housing-at-trio-of-top-dublin-docklands-developments-1.3875456.

Regan, Aidan, and Samuel Brazys. 2018. "Celtic Phoenix or Leprechaun Economics? The Politics of an FDI-Led Growth Model in Europe." *New Political Economy* 23, no. 2: 223–38.

Republic of Ireland. 1993. Dáil Éireann, Ceisteanna, Questions. Oral Answers.—Film Industry. Dublin, April 29. Houses of the Oireachtas. https://www.oireachtas.ie/en/debates/debate/dail/1993-04-29/11/.

Richman-Kenneally, Rhona, and Lucy McDiarmid, eds. 2017. *The Vibrant House: Irish Writing and Domestic Space.* Dublin: Four Courts Press.

Rockett, Kevin, Luke Gibbons, and John Hill. 1987. *Cinema and Ireland.* London: Croom Helm.

Rose, Gillian, Monica Degen, and Clare Melhuish. 2014. "Networks, Interfaces, and Computer-Generated Images: Learning from Digital Visualisations of Urban Redevelopment Projects." *Environment and Planning D: Society and Space* 32, no. 3: 386–403.

Rossiter, Ned. 2016. *Software, Infrastructure, Labor: A Media Theory of Logistical Nightmares.* New York: Routledge.

RTÉ. 2018. "NAMA Cuts Exposure to Ghost Estates from 335 in 2010 to Just 8." January 4. https://www.rte.ie/news/business/2018/0104/931057-nama-2017-review/.

RTÉ News. 2017. "Permission Sought to Appeal Apple Athenry Decision to Supreme Court." December 5. https://www.rte.ie/news/courts/2017/1205/925069-apple-athenry/.

Rubenstein, Michael. 2010. *Public Works: Infrastructure, Irish Modernism, and the Postcolonial.* South Bend, IN: University of Notre Dame Press.

Ruiz, Rafico. 2021. *Slow Disturbance: Infrastructural Mediation on the Settler Colonial Resource Frontier.* Durham, NC: Duke University Press.

Ruuskanen, Esa. 2018. "Encroaching Irish Bogland Frontiers: Science, Policy and Aspirations from the 1770s to the 1840s." In *Histories of Technology, the Environment and Modern Britain,* edited by Jon Agar and Jacob Ward, 22–40. London: University College London Press.

Sadowski, Jathan, and Frank Pasquale. 2015. "The Spectrum of Control: A Social Theory of the Smart City." *First Monday: Peer-Reviewed Journal on the Internet* 20, no. 7. https://papers.ssrn.com/sol3/papers.cfm?abstract_id=2653860.

Sassen, Saskia. 2001 (1991). *The Global City: New York, London, Tokyo,* 2nd ed. Princeton, NJ: Princeton University Press.

Savvas, Antony. 2018. "Dublin to Become Biggest European Data Centre Market in 2019." *Data Economy,* November 6. https://data-economy.com/dublin-to-become-biggest-european-data-centre-market-in-2019/.

Schwab, Klaus. 2015. "The Fourth Industrial Revolution: What It Means and How to Respond." *Foreign Affairs,* December 12. https://www.foreignaffairs.com/articles/2015-12-12/fourth-industrial-revolution.

Scott, Allen J. 2008. *Social Economy of the Metropolis: Cognitive-Cultural Capitalism and the Global Resurgence of Cities.* Oxford: Oxford University Press.

Screen Skills Ireland. n.d. "Who We Are." Accessed July 29, 2020. https://www.screenskillsireland.ie/who-we-are/.

SEAI. 2017. *Ireland's Data Hosting Industry 2017.* https://www.seai.ie/publications/Irelands-Data-Hosting-Industry-2017.pdf.

Searle, Llerena Guiu. 2016. *Landscapes of Accumulation: Real Estate and the Neoliberal Imagination in Contemporary India.* Chicago: University of Chicago Press.

Siggins, Lorna. 2017. "Delight as Apple Decision Heralds 'Silicon Valley of West of Ireland.'" *Irish Times,* October 12. https://www.irishtimes.com/news/ireland/irish-news/delight-as-apple-decision-heralds-silicon-valley-of-west-of-ireland-1.3254139.

Silicon Republic. 2016. "What Makes Ireland the Ultimate Data Centre Capital of Europe?" *Silicon Republic,* March 27. https://www.siliconrepublic.com/enterprise/data-centre-network-ireland.

Simpson, Tim. 2017. *Tourist Utopias: Offshore Islands, Enclave Spaces, and Mobile Imaginaries.* Amsterdam: Amsterdam University Press.

Skvirsky, Salomé Aguilera. 2020. *The Process Genre: Cinema and the Aesthetic of Labor.* Durham, NC: Duke University Press.

Slaby, Alexandra. 2011. "Whither Cultural Policy in Post Celtic Tiger Ireland?" *Canadian Journal of Irish Studies* 37, nos. 1–2: 76–97.

Slattery, Laura. 2016. "Docklands Film Studio Should Go Ahead, Say Poolbeg Backers." *Irish Times,* December 19. https://www.irishtimes.com/business/media-and-marketing/docklands-film-studio-should-go-ahead-say-poolbeg-backers-1.2910553.

Slattery, Laura. 2017. "Creative Ireland Plan Reaching 'Squeaky Bum Time.'" *Irish Times,* May 11. https://www.irishtimes.com/business/media-and-marketing/creative-ireland-plan-reaching-squeaky-bum-time-1.3079690.

Smith, Neil. 2008 (1984). *Uneven Development: Nature, Capital, and the Production of Space,* 3rd ed. Athens: University of Georgia Press.

SONI and EirGrid. 2019. "All-Island Transmission System Performance Report." https://cms.soni.ltd.uk/sites/default/files/media/documents/All-Island-Transmission-System-Performance-Report-2019.pdf.

Srnicek, Nick. 2016. *Platform Capitalism.* New York: Polity Press.

Star, Susan Leigh. 1999. "The Ethnography of Infrastructure." *American Behavioral Scientist* 43, no. 3: 377–91.

Starosielski, Nicole. 2015. *The Undersea Network.* Durham, NC: Duke University Press.

Starosielski, Nicole. 2019. "Internet Infrastructure: Where Foreign Affairs and the Climate Crisis Intersect." *OpenCanada.org,* July 30. https://www.opencanada.org/features/internet-infrastructure-where-foreign-affairs-and-climate-crisis-intersect/.

Stengers, Isabelle. 2017. "Autonomy and the Intrusion of Gaia." *South Atlantic Quarterly* 116, no. 2: 381–400.

Steyerl, Hito. 2013. "How Not to Be Seen: A Fucking Didactic Education .MOV File." *Artforum*. https://www.artforum.com/video/hito-steyerl-how-not-to -be-seen-a-fucking-didactic-educational-mov-file-2013-51651.

Steyerl, Hito. 2017. *Duty Free Art*. Brooklyn: Verso.

Sundaram, Ravi. 2009. *Pirate Modernity: Delhi's Media Urbanism*. New York: Routledge.

Sundaram, Ravi. 2015. "Publicity, Transparency, and the Circulation Engine: The Media Sting in India." *Current Anthropology* 56, supplement: S297–S305.

Swinhoe, Dan. 2022. "EirGrid Says No New Applications for Data Centers in Dublin Until 2028." Data Centre Dynamics, January 11. https://www .datacenterdynamics.com/en/news/eirgrid-says-no-new-applications-for -data-centers-in-dublin-till-2028/.

Tabbara, Mona. 2025. "Irish Post-Production House Windmill Lane Pictures Closes with Immediate Effect." *Screen Daily*, January 10. https:// www.screendaily.com/news/irish-post-production-house-windmill-lane -pictures-closes-with-immediate-effect/5200638.article.

Taffel, Sy. 2023. "Data and Oil: Metaphor, Materiality and Metabolic Rifts." *New Media and Society* 25, no. 5: 980–98.

Tanklar, Allar. 2017. "Investing in Forestry: For the Cash Flow, but Also the Violins and the Poetry." *European Investment Bank*, June 18. http://www.eib .org/en/stories/investing-in-forestry.

Taylor, A. R. E. 2019. "The Data Center as Technological Wilderness." *Culture Machine* 19. https://culturemachine.net/vol-18-the-nature-of-data-centers/ the-data-center-as/.

Teicher, Jordan G. 2014. "A Stroll Through Ireland's Eerie Ghost Estates." *Slate*, August 10. https://slate.com/culture/2014/08/valerie-anex-photographs -ghost-estates-in-ireland-in-her-series-ghost-estates.html.

Terranova, Tiziana. 2000. "Free Labor: Producing Culture for the Digital Economy." *Social Text* 18, no. 2: 33–58.

Thomas, Cónal. 2017. "In a Thriving Corner of the City, Some Small Businesses Are Struggling." *Dublin Inquirer*, October 25. https://www.dublininquirer .com/2017/10/25/in-a-thriving-corner-of-the-city-some-small-businesses -are-struggling.

Thomas, Cónal. 2019. "'A Spotlight on Vacancy': Dublin 8 Locals Gather to Protest Vacant Former Factory." *Thejournal.ie*, January 26. https://www .thejournal.ie/a-spotlight-on-vacancy-dublin-8-locals-gather-to-protest -vacant-former-factory-4458801-Jan2019/.

Thrift, Nigel. 2008. *Non-Representational Theory: Space, Politics, Affect*. New York: Routledge.

Tipton, Gemma. 2009. "Life for the Boom's Dead Spaces." *Irish Times*, June 30, 17.

Titley, Gavan. 2015. "All Aboard the Migration Nation." In *Ireland Under Austerity: Neoliberal Crisis, Neoliberal Solutions,* edited by Colin Coulter and Angela Nagle, 192–215. Manchester: Manchester University Press.

Tom Fleming Creative Consultancy. 2015. *Cultural and Creative Spillovers in Europe: Report on a Preliminary Evidence Review.* October. https://www.artscouncil.org.uk/sites/default/files/download-file/Cultural_creative_spillovers_in_Europe_exec_sum.pdf.

Tovey, Hilary. 1993. "Environmentalism in Ireland: Two Versions of Development and Modernity." *International Sociology* 8, no. 4: 413–30.

Tracy, Tony, and Roddy Flynn. 2016. "Quantifying National Cinema: A Case Study of the Irish Film Board 1993–2013." *Film Studies* 14, no. 1: 32–53.

Tranum, Sam. 2023. "Will the Proposed New Port at Bremore Get Chinese Backing Again This Time Around—and Does It Matter?" *Dublin Inquirer,* November 15. https://dublininquirer.com/2023/11/15/will-the-proposed-new-port-at-bremore-get-chinese-backing-again-this-time-around-and-does-it-matter/.

Trentmann, Frank. 2009. "Materiality in the Future of History: Things, Practices, and Politics." *Journal of British Studies* 48: 238–307.

Tsing, Anna. 2005. *Friction: An Ethnography of Global Connection.* Princeton, NJ: Princeton University Press.

Tsing, Anna. 2009. "Supply Chains and the Human Condition." *Rethinking Marxism: A Journal of Economics, Culture and Society* 21, no. 2: 148–76.

Tsing, Anna. 2015. *The Mushroom at the End of the World: On the Possibility of Life in Capitalist Ruins.* Princeton, NJ: Princeton University Press.

Tubridy, Fiadh. 2023. "Militant Research in the Housing Movement: The Community Action Tenants Union Rent Strike History Project." *Antipode* 56: 1027–46. https://doi.org/10.1111/anti.13014.

Ulin, Julieann Veronica, Heather Edwards, and Sean O'Brien, eds. 2013. *Race and Immigration in the New Ireland.* South Bend, IN: University of Notre Dame Press.

Urry, John. 2014. *Offshoring.* Boston: Polity Press.

Urry, John, and Jonas Larsen. 2011. *The Tourist Gaze 3.0.* New York: Sage.

Van Egeraat, Chris, Sean O'Riain, and Aphra Kerr. 2013. "Social and Spatial Structures of Innovation in the Irish Animation Industry." *European Planning Studies* 21, no. 9: 1437–55.

Velkova, Julia. 2016. "Data That Warms: Waste Heat, Infrastructural Convergence and the Computation Traffic Commodity." *Big Data and Society* 3, no. 2: 1–10.

Velkova, Julia. 2020. "The Art of Guarding the Russian Cloud: Infrastructural Labour in a Yandex Data Centre in Finland." *Digital Icons: Studies in Russian, Eurasian, and Central European New Media* 20: 47–63.

Vonderau, Asta. 2019. "Scaling the Cloud: Making State and Infrastructure in Sweden." *Ethnos: Journal of Anthropology* 84, no. 4: 698–718.

Vonderau, Patrick. 2014. "Industry Proximity." *Media Industries Journal* 1, no. 1: 69–74.

Wall, William. 2011. *Ghost Estate*. Ennistimon, IE: Salmon Poetry.

Wallerstein, Immanuel. 1976. "Semi-Peripheral Countries and the Contemporary World Crisis." *Theory and Society* 3: 461–83.

Ward, Russ, and John Carcone. 2015. "A New Approach to Financing Your Data Center." *Sirius Edge*, April 3. https://edge.siriuscom.com/infrastructure-operations/a-new-approach-to-financing-your-data-center.

Watkins, James. 2017. "Fintech Companies Vie for Some of That Irish Luck." *Ozy*, January 4. https://www.ozy.com/around-the-world/fintech-companies-vie-for-some-of-that-irish-luck/74576/.

Watts, Laura. 2018. *Energy at the End of the World: An Orkney Islands Saga*. Cambridge, MA: MIT Press.

White, Pauline. 2010. "Creative Industries in a Rural Region: *Creative West*: The Creative Sector in the Western Region of Ireland." *Creative Industries Journal* 3, no. 1: 79–88.

Whiting, Sam, Tully Barnett, and Justin O'Connor. 2022. "'Creative City' R.I.P.?" *M/C Journal* 25, no. 3. https://doi.org/10.5204/mcj.2901.

Woods, Killian. 2020. "Data Centres Use Same Amount of Water as Large Towns." *Business Post*, June 14. https://www.businesspost.ie/utilities/data-centres-use-same-amount-of-water-as-large-towns-b1092219.

Woods, Michael. 2007. "Engaging the Global Countryside: Globalization, Hybridity and the Reconstitution of Rural Place." *Progress in Human Geography* 31, no. 4: 485–507.

WRAP Fund. n.d. "Production." Accessed July 29, 2020. http://wrapfund.ie/wrap-funding/production/.

Wuerthele, Mike. 2018. "Apple's $1 Billion Athenry, Ireland Data Center Approved After Legal Challenges Squashed." *Apple Insider*, October 12. https://appleinsider.com/articles/17/10/12/apples-1-billion-athenry-ireland-data-center-approved-after-legal-challenges-squashed.

Yusoff, Kathryn. 2018. *A Billion Black Anthropocenes or None*. Minneapolis: University of Minnesota Press.

Zayo. 2017. "Spotlight on Dublin: The World's Data Centre." May 30. https://www.zayo.com/uk/spotlight-dublin-worlds-data-centre/.

INDEX

Page references in italic represent illustrations.

climatic politics, of data centers, 7, 167
cloud infrastructure, 60, 173, 191, 192
"cloud ruins," 205
Clover, Joshua, 115
cognitive capitalism, 92
cognitive-cultural capitalism, 87, 90
Coillte, 29, 59, 108, 197, 199–200; Apple and, 154, 160, 187, 199, 243n13
colonial geography, 106, 192
colonization: anti-colonial liberation, 221; anti-colonial solidarity, 231n11; colonial division of the world, 26; Global South, postcolonial states in, 40, 48; postcolonial dysphoria, 83; postcolonial Irish state, 5, 12, 13, 36, 37, 38–39, 42, 58; postcolonial residues of power, 58; precolonial heritage, 26, 55; resource mapping, 5, 230n6
Commission for Regulation of Utilities (CRU), 242n10
commodification, Ireland and, 14, 27, 55, 111, 208
Community Action Tenants Union (CATU), 236n7
commuters, 58, 187, 209, 211, 242n8
compulsory purchase orders, 184
computer components, 65, 69
Connelly, Charlie, 190–91
Connemara, County Galway, 153, 154
Constitution of Ireland, Amendments to, 13
contemporary capitalism, 15, 25, 51, 68, 72, 171, 178, 200, 213
contract work, 33, 50, 118, 119, 120, 132, 157. See also freelance workers
Cook, Tim, 189
Cooper, Melinda, 44, 200
coproduction agreements, 130–31
Cork City, 169, 175
Cork Internet Xchange (CIX), 169, 171, 175, 193
Corporation Tax: Apple, back taxes and, 189; decline in income, 209, 210; Ireland, rate in, 40, 41, 42, 87, 91, 192, 193; OECD tax harmonization, 233n7, 243n19
Costello, Dan, 192
Coulter, Colin, 48
Court of Appeal, 187, 189
Covanta, 78

COVID-19 pandemic, 217, 218, 236n10
Cowen, Deborah, 28, 66, 69, 106
creative city, 24, 52, 57, 63, 90; built environment in, 16, 92, 93, 99; constant construction of, 94, 97; creatives living in, 89; DCC's planning policies, 87–88; Dublin as, 7, 32, 76, 77, 78–79, 86, 87; Galway city, UNESCO-designated, 132; mental and physical space, blurring of, 114; visual narration of, capitalists and, 99, 106–7. See also Dublin Docklands; Limerick City; Poolbeg Peninsula; Poolbeg West SDZ
"creative class" thesis, 87, 89, 91
Creative Europe, 128
"creative fix," 12, 73, 91, 92, 115
creative hubs, 89, 145, 237n20
creative industries, 31, 32, 33, 52, 53–54, 87–88, 89–90; blue-collar labor, exclusion of, 89; capital and, 86; Dublin Docklands and, 87; Limerick and, 23; media policy and, 118; on- and off-set class hierarchy in, 148; Pigeon House Power Station and Hotel site, 110; precarity, normalization of, 56, 92, 147; rural areas and, 142–43. See also creative workers; Irish workers; media workers
Creative Ireland, 88, 140
Creative Limerick, 138, 240n17
creatives, 10, 89, 92, 119, 120, 124
creative spaces, privatization of, 114–15
Creative Spaces Collective, 92, 109–10
Creative West report, 142, 145
creative workers, 32, 52, 88, 90, 122, 142; classification of, 148; Dingle Peninsula and, 145, 147; in the Global South, 129
"crisis ordinary," 81, 199, 202, 211–12
cultural diplomacy, 129–30, 198
cultural enterprise, 126, 128
cultural groups, 92
cultural identity, 11; 130–31, 152, 231n12
cultural policy, 11, 56, 88, 130; postfinancial crisis, 128; service industries and, 138; tourism and, 138, 142

www.ingramcontent.com/pod-product-compliance
Lightning Source LLC
Chambersburg PA
CBHW020314290526
45785CB00007B/2785